Technological Nature

Technological Nature

Adaptation and the Future of Human Life

Peter H. Kahn, Jr.

The MIT Press
Cambridge, Massachusetts
London, England

For information about special quantity discounts, please email special_sales@mitpress.mit.edu

This book was set in Sabon by Toppan Best-set Premedia Limited. Printed and bound in the United States of America. Printed on recycled paper.

Library of Congress Cataloging-in-Publication Data

Kahn, Peter H.
Technological nature : adaptation and the future of human life /
Peter H. Kahn, Jr.
    p.   cm.
Includes bibliographical references and index.
ISBN 978-0-262-11322-9 (hardcover : alk. paper)
1. Technological forecasting. 2. Technology—Social aspects. 3. Bionics.
I. Title.
T174.K35   2011
303.48′3—dc22
                                                    2010035834

10   9   8   7   6   5   4   3   2   1

Contents

Acknowledgments

I like trees and people, difficult intellectual discussions, and good research. Thus I have been fortunate to be part of an extraordinarily talented group—the Human Interaction with Nature and Technological Systems (HINTS) Laboratory—at the University of Washington. Current and recent members include Brian Gill, Nathan Freier, Rachel Severson, Jolina Ruckert, Aimee Reichert, Solace Shen, Heather Gary, John Lin, Cady Stanton, Nicole Kennerly, and Lorin Dole. Earlier members also include Jennifer Abdo, Irene Alexander, Erin Fowler, Amanda Ryan, Brandon Rich, Jeff Anderson, Louise Wun Choi, Jesse McPherron, Scott Santens, and Jonathan Sabo.

Thanks, too, to many members of the Value Sensitive Design Group, directed by Batya Friedman, and including Alan Borning, Lisa Nathan, Janet Davis, Shaun Kane, and Pedja Klasnja.

Brian Gill directed most of the statistical analyses that underlie the empirical results I report on in chapters 4 through 10. Mike Eisenberg (Dean Emeritus of the Information School at the University of Washington) solved numerous problems that enabled the research. Solace Shen assisted ably with manuscript preparation.

My daughter, Zoe Kahn, helped me think through some of the ideas and offered practical advice.

Barbara Dean provided helpful comments on many of the chapters, and for decades we have discussed many of the ideas of this book and more.

I extend thanks to other colleagues for our long-standing intellectual discussions, including Gene Myers, Carol Saunders, Hiroshi Ishiguro, Takayuki Kanda, Gail Melson, Alan Beck, Karl MacDorman, Cecilia Wainryb, Charles Helwig, Judi Smetana, and Larry Nucci. Thanks to

Melanie Killen for encouraging my research when it was emerging, and when its contributions were not as readily apparent to others. Elliot Turiel was my graduate advisor and mentor at the University of California, Berkeley, during the 1980s. His influence can still be felt throughout much of this book, and it is still appreciated. Thanks to my editor at MIT Press, Clay Morgan. This is my third book with him.

Patricia Hasbach provided insights into several of the difficult conceptual chapters of this book within the context of our new project on the rewilding of the human species.

Thanks to Ephraim Glinert, William Bainbridge, and Suzi Iacono at the National Science Foundation. The research reported in this book has been supported by the National Science Foundation under grant nos. IIS-0102558 and IIS-0325035. Any opinions, findings, and conclusions or recommendations expressed in this book are mine and do not necessarily reflect the views of the National Science Foundation.

My empirical research is collaborative, as research of this scope often is. I extend thanks to my collaborators and to others who assisted with the research reported in the following empirical chapters.

A Room with a Technological Nature View (Chapter 4)

My collaborators were Batya Friedman, Brian Gill, Jennifer Hagman, Rachel L. Severson, Nathan G. Freier, Erika N. Feldman, Sybil Carrère, and Anna Stolyar. Judith H. Heerwagen, Gordon H. Orians, and James A. Wise were involved in early discussions about this research.

Office Window of the Future? (Chapter 5)

Batya Friedman was the lead author. My other collaborators were Nathan G. Freier, Peyina Lin, and Robin Sodeman.

Hardware Companions? (Chapter 6)

My collaborators were Batya Friedman and Jennifer Hagman. Kathleen Crosman and Joseph Goldberg assisted with data entry, and Ann Hendrickson assisted with reliability coding. Pamela Hinds and Sara

Kiesler provided substantive comments on an earlier version of this reported research.

Robotic Dogs in the Lives of Preschool Children (Chapter 7)

My collaborators were Batya Friedman, Deanne R. Pérez-Granados, and Nathan G. Freier. Jeff Anderson, Norma Barajas, Adrian Guzman, Annie Hendrickson, Josh Kienitz, Ilene Lewis, Sandra Yu Okita, Rachel L. Severson, and Tyler Stevens assisted with data collection. This research was funded, in part, by (a) a grant from the University of Washington's Center for Mind, Brain, and Learning and the Talaris Research Institute and Apex Foundation, the family foundation of Bruce and Jolene McCaw to P. H. Kahn, Jr., and B. Friedman, and (b) a gift from Intel Corporation to B. Friedman and the Information School at the University of Washington.

Robotic Dogs and Their Biological Counterparts (Chapter 8)

Gail F. Melson was the lead author. My other collaborators were Alan Beck, Batya Friedman, Trace Roberts, Erik Garrett, and Brian Gill. Brian Gilbert, Migume Inoue, Oana Georgescu, and Jocelyne Albert assisted with data collection, transcript preparation, and coding.

Robotic Dogs Might Aid in the Social Development of Children with Autism (Chapter 9)

Cady M. Stanton was the lead author. My other collaborators were Rachel L. Severson, Jolina H. Ruckert, and Brian T. Gill. Advanced Telecommunications Research, and specifically Norihiro Hagita, Hiroshi Ishiguro, and Takayuki Kanda, provided use of their debugger software system, which we used for reviewing video footage and coding data. Earl Wilson was involved in discussions about autism at various phases in this research. Thanks, too, to Daniel Glick, Heidi Shaw, and students at Yakima Valley Community College and Children's Village and the Autism Resource Team of Yakima, Washington. Along with support from the National Science Foundation, acknowledged earlier, this research was supported, in part, by a University of Washington Mary Gates Undergraduate Research Scholarship to Cady M. Stanton.

The Telegarden (Chapter 10)

My collaborators were Batya Friedman, Irene S. Alexander, Nathan G. Freier, and Stephanie L. Collett. Thanks to Ken Goldberg and Hannes Mayer for providing us with the Telegarden chat data. Brian Gill, Erika N. Feldman, Jonathan Sabo, and Nicole Gustine assisted with aspects of this project.

Some of the material draws from my previously published peer-reviewed papers, listed below. The material is reprinted with permission from the publishers.

References

Friedman, B., N. G. Freier, P. H. Kahn, Jr., P. Lin, and R. Sodeman. 2008. Office window of the future?—Field-based analyses of a new use of a large display. *International Journal of Human-Computer Studies, 66,* 452–465.

Friedman, B., P. H. Kahn, Jr., and J. Hagman. 2003. Hardware companions?—What online AIBO discussion forums reveal about the human-robotic relationship. In *Proceedings of the Conference on Human Factors in Computer Systems* (pp. 273–280). New York: ACM Press.

Kahn, P. H., Jr., B. Friedman, I. S. Alexander, N. G. Freier, and S. L. Collett. 2005. The distant gardener: What conversations in the Telegarden reveal about human-telerobotic interaction. In *Proceedings of the IEEE International Workshop on Robot and Human Interactive Communication* (pp. 13–18). Piscataway, NJ: IEEE.

Kahn, P. H., Jr., B. Friedman, B. Gill, J. Hagman, R. L. Severson, N. G. Freier, et al. 2008. A plasma display window?—The shifting baseline problem in a technologically-mediated natural world. *Journal of Environmental Psychology, 28,* 192–199.

Kahn, P. H., Jr., B. Friedman, D. R. Pérez-Granados, and N. G. Freier. 2006. Robotic pets in the lives of preschool children. *Interaction Studies: Social Behavior and Communication in Biological and Artificial Systems, 7,* 405–436.

Kahn, P. H., Jr., R. L. Severson, and J. H. Ruckert. 2009. The human relation with nature and technological nature. *Current Directions in Psychological Science, 18,* 37–42.

Melson, G. F., P. H. Kahn, Jr., A. Beck, and B. Friedman. 2009. Robotic pets in human lives: Implications for the human–animal bond and for human relationships with personified technologies. *Journal of Social Issues, 65,* 545–567.

Melson, G. F., P. H. Kahn, Jr., A. Beck, B. Friedman, T. Roberts, E. Garrett, et al. 2009. Children's behavior toward and understanding of robotic and living dogs. *Journal of Applied Developmental Psychology, 30,* 92–102.

Stanton, C. M., P. H. Kahn, Jr., R. L. Severson, J. H. Ruckert, and B. T. Gill. 2008. Robotic animals might aid in the social development of children with autism. In *Proceedings of the ACM/IEEE International Conference on Human-Robot Interaction* (pp. 97–104). New York: ACM Press.

Introduction

In his treatise, *The Embers and the Stars: A Philosophical Inquiry into the Moral Sense of Nature*, Erazim Kohak (1984) writes:

> Though philosophy must do much else as well, it must, initially, see and, thereafter, ground its speculation ever anew in seeing. So I have sought to see clearly and to articulate faithfully the moral sense of nature and of being human therein through the seasons lived in the solitude of the forest, beyond the powerline and the paved road, where the dusk comes softly and there still is night, pure between the glowing embers and the distant stars. (p. xii)

It is possible that most of us have had this experience—of the waning campfire, and then so softly and so gently the darkness of the night sky takes hold. It is calming, simple, and awe-inspiring all at the same time. Such experiences with nature go back hundreds of thousands of years in our evolutionary history.

Yet, now, in recent years—especially given the sophistication and pervasiveness of our computational technologies—we have begun to change our species' long-standing experiences with nature. Now we have what I am calling *technological nature*: technologies that in various ways mediate, augment, or simulate the natural world.

Entire television networks, such as Discovery Channel and Animal Planet, provide us with mediated digital experiences of nature: the lion's hunt, the monarch butterfly's migration, and the adventure of climbing high into the Himalayan peaks. Video games like Zoo Tycoon engage children with animal life. Zoos themselves are bringing technologies such as webcams into their exhibits so that we can, for example, watch the animals in their captive habitat from the leisure of our homes. Inexpensive robotic pets, such as the i-Cybie, Tekno, and Poo-Chi, have been big sellers in Walmart and Target. Sony's higher-end robotic dog AIBO portends the future. A few years ago, you could telehunt in Texas. You would go online from your computer in New York City or Amsterdam, or from an Internet café in Mumbai, or anywhere on this planet and control a mounted rifle through a Web interface and hunt and kill a live animal. The animal would then be gutted and skinned by the owner of the establishment, and the meat shipped to your doorstep. Texas outlawed telehunting, but teleshooting still exists, using targets instead of animals. I recently toured Microsoft's "home of the future," a physical building at their Redmond, Washington, facilities. One of their technological prototypes included a personified agent for the home, "Grace," who controlled many of the home's functions and communicated through a voice interface. Toward the end of the tour, my Microsoft guide said, "Grace, it's nighttime." Just like that the lights dimmed, and projected on the walls was the night sky. If we coupled that projection with the DVD of the crackling wood fire in the hearth, we would give new meaning to Kohak's evocation of the "night, pure between the glowing embers and the distant stars."

In terms of the future well-being of our species, does it matter that we are replacing actual nature with technological nature? This book seeks answers to that question.

Over the last seven years, my colleagues and I have investigated children's and adults' experience of different forms of cutting-edge technological nature. We created, for example, a "technological nature window" by installing an HDTV (high-definition television) camera on top of a building on our university campus and then displaying a real-time local nature view on fifty-inch plasma screen "windows." In an experimental study, we compared the physiological and psychological effects of experiencing the technological nature window view to those of experiencing a glass window view of the same scene and no view at all. In a sixteen-week field study, we deployed these technological nature windows in a handful of "inside" (no-window) offices of university faculty and staff. In four other studies, we investigated people's experiences interacting with the robotic dog AIBO. The first study investigated how people conceptualized AIBO through their spontaneous written discourse in online AIBO discussion forums. The second study investigated preschool children's behavior with and reasoning about AIBO compared to a stuffed dog. The third study investigated older children's and adolescents' behavior with and reasoning about AIBO compared to a biologically live dog. And the fourth study investigated whether interacting with AIBO could benefit the social development of children with autism. We also conducted a study that involved telepresence, but not by employing telehunting or teleshooting, as described above, but a Telegarden: a garden in Austria created by Ken Goldberg and his colleagues that allowed remote computer users to plant and tend seeds by controlling a robotic arm through a Web-based interface.

These studies form the core empirical base of this book (chapters 4–10). Through these studies—and an account of our ancestral origins as a natural, technological, and adapting species—I ask the following questions:

• What are the physiological, emotional, and cognitive effects of interacting with technological nature? How do such effects compare to experiencing actual nature and experiencing no nature?

• What are the benefits and limitations of experiencing technological nature?

• Are the limitations of technological nature due to the current state of the technology, or is there something fundamentally limiting about it such that it will always come up short in terms of the human experience?

• How can we design technological nature to enhance human well-being?

• Through interaction with personified technological nature (e.g., animal robots), might children construct an entirely new ontological category —a category of existence, for example, that is neither alive nor not alive, but both?

• Over the span of a single lifetime, multiple generations, and 10,000 years, how will humans adapt to technological nature?

• Can we as a species flourish apart from actual nature?

One goal of this book is to motivate the above questions, and to show how they are tractable from an empirical, scientific standpoint. The second goal is to provide partial answers. The third goal is to provide a theoretical framework for further investigations, grounded in evolutionary and psychological theory, ethically shaped, and with an eye toward the future of our species.

In reporting my data, I use descriptive statistics (usually percentages) and highlight findings that are statistically significant based on inferential statistics. But I steer clear of statistical details, which I have reported formally in peer-reviewed scientific journal articles. When appropriate, I cite these articles and ask the more statistically interested reader to consult them.

For the reader who is looking for a quick summary of my chapters, I have provided short conclusions at the end of each. Together the conclusions might take twenty minutes to read. If you spend that time, you would have a concise summary of the book.

Even more briefly, here is how I see things. Two world trends are powerfully reshaping human existence. One is the degradation and destruction of large parts of the natural world. A second is unprecedented technological development, in terms of its sophistication and pervasiveness. At the nexus of these two trends lies technological nature. My question is: Does it matter for the physical and psychological well-being of the human species that actual nature is being replaced with technological nature? I think the answer is yes.

One of my canonical pairs of studies is with the technological nature window. My colleagues and I found (chapter 4) that in terms of heart rate recovery from low-level stress, working in an office environment with a glass window that looked out on a nature scene was more restorative than working in the same office without the outside nature view (in effect, an inside office). Nature is good for us. Second, in terms of this same physiological measure, the technological nature window was no different from the inside office. In other words, in terms of stress reduction, it appears likely that an actual nature view is better than a technological nature view. That does not mean, however, that there are no benefits to a technological nature window. There are. We found (chapter 5), for example, that people who worked long-term in an inside office with a technological nature window reported benefits from its use. This general pattern—that technological nature is better than no nature but not as good as actual nature—held up across the studies using our two other technological platforms: robotic pets and a Telegarden.

If we employed technological nature only as a bonus on top of our interactions with actual nature, then we would be in good shape.

Unfortunately, we keep degrading and destroying actual nature, and we are becoming increasingly impoverished for it.

This trend is difficult to reverse because it is hard for humans to believe that it is even happening. For example, if you try to explain what we, as humans, are missing in terms of the fullness of the human relation with nature, a well-meaning person can look at you blankly—it has happened to me many times—and respond "but I don't think we're missing much." We are missing plenty. Why do we not know? A large part of the reason likely involves what in my earlier work I termed *environmental generational amnesia*. The basic idea here is that each generation constructs a conception of what is environmentally normal based on the natural world encountered in childhood. With each generation the amount of environmental degradation increases, but each generation tends to take that degraded condition as the nondegraded condition, as the normal experience. In chapter 11, I expand on this idea and bring forth further evidence and argument. It is hard enough to solve environmental problems, such as global climate change, when we are aware of them; it is all the harder when we are not. Thus I believe that the problem of environmental generational amnesia will emerge as one of the central psychological problems of our lifetime.

In this book, my voice is as scientist, environmentalist, technologist, and humanist. This integration of voices might at times puzzle the reader, but it is part of the package I bring forward. I write of nature as gentle, nurturing, deadly, and that it makes us alive. I believe in science yet reject deterministic and mechanistic conceptions of mind, self, and society. I do not draw on the philosophical proposition that nature has inherent value or moral standing independent of human life, though that may be true. Rather, my position is decisively human centered. I argue that as a species our fulfillment and flourishing lies not in dominating nature but in living in relation with it, grounded in patterns of interactions with nature that extend far back in our evolutionary history, and which shape us still. We are a natural species. We are also a technological species. We love technology, and it has been adaptive, too. I love technology. I am not a Luddite. But I am also keenly aware that there are costs that accompany almost every technological innovation. Thus one challenge, which I take up in this book, is of how to design technological systems in general, and technological nature specifically, for more human and humane purposes.

Some people say that because we are an adapting species, we do not have to worry about the long-term effects of the loss of nature and the increase of technological nature. I argue in chapter 12 that that is a mistake. Even if this planet over the next hundred years (or let us imagine a thousand years or ten thousand years) could sustainably support 6 or 9 billion people as biologically living beings, it cannot sustainably support even one-tenth that number in ways where humans fully flourish in their relationship with nature. Almost all discussions people and societies at large have about environmental issues unduly focus—explicitly or implicitly as the bottom line—on how the issues affect our health, our income, and our material possessions. These matters are important, of course. But what about that which, in our relation with nature, gives life further meaning? Consider a caged elephant in a zoo. It survives for years in the space of a small parking lot, while its biological programming and its ancestral self wants and needs the wild and vast spaces of its origins. We are like animals in a zoo. We are caging ourselves. There are millions of children who have never slept out under the stars. There are millions of children who have never in their lives seen the stars, because of the air pollution and light pollution in their cities. Can you imagine that? Can you imagine growing up and never having seen a star? As I write in chapter 11, none of us has experienced the plethora of winged-life of the now-extinct passenger pigeons or, as Lewis and Clark wrote, of "10 thousand buffaloe within a circle of 2 miles around" (Moulton 1993, p. 106).

What I am saying in this book is that humans flourishing matters just as much as humans existing. Toward that end we need to reenvision what is beautiful and fulfilling and often wild in essence in our relationship with nature.

1

The Old Way

As a species, we love nature. It could not have been otherwise. For our minds and bodies came of age hundreds of thousands of years ago on the African savannahs, where certain patterns of interaction with nature contributed to our survival and psychological well-being.

Back then, for example, we sought out bodies of water, hunted, dug deep for roots, walked the land when our predators did not, and situated ourselves with landscape that could offer prospect and refuge. There were hundreds of such interaction patterns with nature. These patterns are with us still. They comprise part of our essential selves. But in our current technological world, they can be difficult to see, and they can sometimes take perverse forms. If we go to a zoo, for example, we can sometimes see a child, or even adult, throwing food or a pebble at an animal, such as a lion, leopard, or great ape—even when the signage says not to. The person is trying to get the animal's attention. Why? Perhaps it is because for the entire history of our species we have not only been aware of wild animals, but we have been aware that they have been aware of us, and the desire for that form of interaction persists in modern times. But how would we know that to be true? How would we know how people interacted with nature so long ago?

One answer is to draw on accounts from the 1950s of indigenous African people, the Bushmen, before they had virtually any contact with the West. The Bushmen had lived sustainably in the African landscape for about 35,000 years, with a lifestyle that presumably changed little during that time. Thus, to understand them then, in the 1950s, is to gain access, in a way that bones and fossils cannot offer, to aspects of who we were at an earlier time in our evolutionary history.

In 1950, the Marshall family set out into the Kalahari Desert to find and live with the Bushmen. It was hard going in their vehicles. Their tires punctured, springs broke, and grass seeds clogged the radiators, causing them to overheat, boil over, and waste precious water. The trucks sometimes sank up to their axles into the soft sand. But they eventually reached a waterhole in a spot called /Gam that had in years past been used by Bushmen. The waterhole had been usurped by an absentee Tswana rancher who grazed about twenty head of cattle in the area. The cattle were managed by a Tswana family. While the Marshalls were camping in /Gam, two Bushmen walked into their camp. They were heading deeper into the Kalahari and agreed to guide them further. Through this means, the Marshalls eventually made contact with the Ju/wasi—one of the five groups of Bushmen that inhabited the Kalahari. Between 1950 and 1955, the Marshalls made three expeditions into this area and spent a total of two years living with the Bushmen. The Marshall family consisted of the father (Laurence), mother (Lorna), son (John), and daughter (Elizabeth). The daughter at the time of the first visit was 19 years old. Based on her two years of living with the Ju/wasi, she wrote an initial book, *The Harmless People* (Thomas 1959), and a more recent one, *The Old Way* (Thomas 2006). After 1955, Laurence, Lorna, and John Marshall lived many more years with the Bushmen, and Lorna Marshall (1976) wrote *The !Kung of Nyae Nyae*.

Based on the Marshall family's experiences and writings, I would like to draw out six broad interaction patterns that help us understand the Ju/wasi relationship with nature. As I mentioned above, I think these forms of interaction comprise part of our essential selves; and I will come back to them later (chapter 9) to help frame an account of how humans can flourish with technological nature.

Using Bodies, Vigorously

To sustain their life, the Ju/wasi had to exert themselves physically. Food gathering, for example, took effort. On occasions, according to Thomas (2006), the men and women would visit a mangetti grove and each carry home from fifty to one hundred pounds of nuts. Thomas describes one day foraging with women that included six miles of walking to where edible roots grew. Each woman found and dug for her own roots. The roots lay several feet deep, often weighed between five and fifteen pounds, and would take as long as fifteen minutes to dig. The women spent the rest of the daylight digging roots. Near sunset, they began the long walk home. Some of the women were nursing their babies, and carried them, too. Other women had young children who could walk but eventually tired, and needed to be carried. Thus many of the women carried their roots and child, as much as eighty pounds. The women themselves weighed about ninety pounds. Thomas estimates that Ju/wa women walked about 1,500 miles a year.

Hunting was physically difficult, as well. Giraffes, elands, kudus, wildebeests, gemsboks, and hartebeests were the animals of choice. Bow and poison arrow were the usual means. The hunters, always men, would head out in the direction they thought promising. Upon finding a heard of wildebeests, the hunters would stalk close, and then let fly with poison arrows. An arrow itself did not kill; it was simply the means to get the poison into the animal's bloodstream. The wildebeests would bolt. At first, the victim would not feel sick. After a day, as the poison did its work, the animal would weaken. "All this time, the hunters would be following, stopping in the late afternoon to make a fire and wait for morning, picking up the trail the next day" (Thomas 2006, p. 100). After two or three days, the hunters might find the animal "dead, or standing alone in some bushes, trying to hide, or lying down, too weak to stand" (ibid.). If the latter, they would kill the animal with their spears. They skinned and butchered the animal, and each carried heavy loads of meat home. The hunters often traveled very far; the conditions could be very harsh. Marshall (1976) writes that during the hottest parts of the year, the temperature of the surface sands reached 140 degrees Fahrenheit. During the middle parts of such days, the hunters would "seek shade, scrape out a shallow hollow, urinate in it, lie down on the moistened place, and dust a light layer of sand over themselves. They wait like this till the day cools, minimizing dehydration and saving their feet" (ibid., p. 68). Another method of hunting the eland was equally demanding. The mature bull eland, in particular, with large amounts of body mass, could be overcome on especially scorching days by a runner who kept at him mile after mile. The runner could not match the eland for speed, for the eland sprints at 35 miles an hour. But eventually, after many hours of being chased in the heat, the eland overheats and can run no more. "Then the hunter, with the last of his strength, can catch up and grab him by the tail, then kill him with a spear if he brought one, or he can push the eland over and lie on his neck to keep him from struggling and clamp his hands over the eland's nose and mouth to stop his laboring breath" (Thomas 2006, p. 32).

Periodicity in the Satisfaction of Physical Needs and of Nature

The Ju/wasi experienced wide variation in a scarce landscape. After a successful hunt, for example, there could be large quantities of meat for several days. Other times there would be very little. Marshall writes: "The intense craving for meat, the uncertainty and anxiety that attend the hunt, the deep excitement of the kill, and finally, the eating and the satisfaction engage powerful emotions in the people" (as quoted in Thomas 2006, p. 102). Water was usually scarce. When hunters set out, "they would drink prodigious amounts of water, until their stomachs were bulging, and would do this again on returning, and in between would choose water over food if the two sources were not together" (Thomas 2006, p. 95). The Ju/wasi lived in a desert with few waterholes. Still, when the rains came, there would be brief times when the desert pans filled with water and the children would swim. There were times of activity, of hunting and foraging, of course; but these were balanced by times of leisure around camp: storytelling around a fire while food is being cooked; beading with ostrich eggshells; playing music on their hunting bows. The Ju/wa lives were responsive to sunrise and sunset, the migration of animals, when berries would be sweet, roots ready for digging and mangetti nuts ready for harvesting, clouds forming, seasons changing, heat and cold, predators resting.

Freedom of Movement

The Ju/wasi were a nomadic people. This does not mean that they moved camp every day. Far from it. But movement often solved the problems they faced. "When, for instance, a certain kind of biting maggot infested the shelters of one Ju/wa group, those people moved. They didn't go far, but they moved" (Thomas 2006, p. 161). When waterholes dried up, food supplies dwindled, or tempers flared, they moved. Their entire cultural system "was based upon the ability to relocate" (ibid.). How did they move so easily? Thomas writes:

We had tents, cots, sleeping bags, folding chairs and tables, maps, a compass, cameras, film, recording equipment, reference books, notebooks, pens, ink, pencils, disinfectants, antivenin kits for snakebites, brandy, cases of canned foods, boxes of dry foods, dishes, cooking pots, frying pans, knives, forks, spoons, cigarettes, matches, spare tires, auto parts, inner tubes, tire patches, jacks, toolboxes, winches, motor oil, drums of gasoline, drums of water, bars of yellow soap, towels, washcloths, toothpaste, toothbrushes, coats, sweaters, pants, boots, sneakers, shirts, underwear, socks, reading glasses, safety pins, scissors, a sewing kit, binoculars, bullets, a rifle.

The Ju/wasi had sticks, skins, eggshells, grass. (p. 62)

Environmental Checks and Balances

Thomas (2006) suggests that the Bushmen of the Kalahari represent "the most successful culture that our kind has ever known, if . . . stability and longevity are measures" (p. 6). Thirty-five thousand years is a long time for a people to inhabit an area. Perhaps part of their success lay in their limited technological prowess. They had digging sticks, bows and arrows, and spears. Digging sticks were also used to balance a load, as a lever and cane, to knock nuts from a tree, and even on occasion to fend off an attacking predator. There was not much else in terms of their tool use. Thus their practices were strongly shaped and checked by a highly demanding environment.

A compelling example is of family size. Most couples had from one to four living children. Yet they had no mechanical or medical methods of birth control. How were these limits achieved? In part they were achieved physiologically: "The strenuous work and absence of body fat prevented hunter-gatherer women from menstruating at an early age, and after the burden of lactation was added to their bodies, they did not menstruate nearly as often as do the women of agricultural and industrial societies. They certainly did not menstruate monthly" (Thomas 2006, p. 192). Once with child, a woman would breastfeed for about four years. Thomas writes that if "Ju/wa children of the Old Way were less than three or four years apart, both children would be undernourished and both could die" (ibid., p. 196). Even if it was possible to keep both young children nourished, their nomadic lifestyle allowed for a mother to carry her belongings and one child; but to carry two young children would be physically impossible. Thus, when necessary, and it was not often, limits were also achieved by the mother practicing infanticide. As Thomas writes: "Although lactation, strenuous work, and a low-fat diet almost always prevented conception, on very rare occasions a woman conceived anyway, even though she had a nursing baby under the age of three or four, and when this happened she might have to dispose of the new baby at birth. . . . The act was called //kao, which means 'throw down,' 'hurt,' or 'go from'" (ibid.). It was a sad time for the mother, and for the group. But everyone recognized that there was no choice.

Had there been a choice, had the mother the technological means to nourish and physically carry two babies at the same time, and had many mothers the same means, the Ju/wa group size would have grown rapidly. But then the desert's food and water supply would have been insufficient and a large social calamity would have eventually ensued. Even at their current size, there were some age gaps: "a group might have knee children and babies, but then no other children younger than their teens. This, too, could suggest a time of hunger, when the children who would have filled that gap had died of starvation as babies" (Thomas 2006, p. 195).

Encountering the Wild

The Kalahari Desert might have initially looked wild to the Marshall family, if by wild we mean strange and unknown. But it was not wild in this way to the Ju/wasi. Thomas (2006) believes that most if not all of the land of Nyae Nyae was known to the Ju/wasi, an area covering about six thousand square miles, and that all of the existing watering holes were known: seven holes the Ju/wasi considered permanent and eight semipermanent, which could go dry during drought years. And, of course, the Ju/wasi were psychologically comfortable in their landscape. It was their home.

But still the Ju/wasi encountered wildness in other forms. The most obvious was with predators, especially leopards, hyenas, and lions. "One night in Nyae Nyae, a leopard crept up behind a man who was sitting with his wife at their fire. The leopard seized the man from behind, at the nape of the neck. His wife leaped to her feet, snatched up her digging stick, and whacked the leopard so hard on the head that he let go of his victim. . . . Despite their preference for primates, leopards are easily discouraged" (Thomas 2006, p. 139). Hyenas occasionally could come into a camp and take bites out of sleeping people—that is their style, to get mouthfuls of flesh and not to kill first and then eat second. Lions could catch and take down what they wanted. "No other creatures of the savannah sleep as deeply or as soundly as lions, but after all, lions are the main reason for not sleeping soundly" (ibid., p. 148). Lions are also curious about their surroundings. Sometimes at night they walked into Ju/wa camps and poked around. Thomas writes of the following encounter:

I remember one night in particular. . . . I soon heard a man's strong voice in a stern, commanding tone telling someone to leave immediately. The Ju/wasi never took that tone with one another. I came out of the tent to see what was happening, and behind some of the shelters I saw four very large lions, each three times the size of a person. . . . The speaker was ≠Toma. Without taking his eyes off the lions, he repeated his command while reaching one hand back to grasp a flaming branch that someone behind him was handing to him. He slowly raised it shoulder-high and shook it. Sparks showered down around him. "Old lions," he was saying firmly and clearly, "you can't be here. If you come nearer we will hurt you. So go now! Go!" . . . The lions watched ≠Toma for a moment longer, then gracefully they turned and vanished into the night. (pp. 150–151)

Wildness can also refer to states that are vast, free, and self-organizing. That is a description of nature, unencumbered and unmediated by technological artifice. In this sense of the term, the Ju/wasi encountered wildness on a daily basis. Wildness did not just exist in facing off with a lion while brandishing a burning branch. It was encountered in the periodicity described above: chasing down a bull eland in 120 degree heat; experiencing the migration of birds, the changing of the seasons, heat and cold. It was the freedom to move, and the strength to do so, and the land to do it in.

Cohabiting with Nature

Both the lions and the Ju/wasi lived by the watering holes. Lions are dangerous predators. But very rarely did lions attack the Ju/wa people. Why? Thomas (2006) offers an explanation that fills out a conception of what it means for a people to cohabitate with nature.

Part of the Old Way, according to Thomas (2006), entails avoiding conflict whenever possible. Hyenas and lions, for example, visited the watering holes at different times of the night, thus avoiding one another and avoiding conflict. The Ju/wasi followed the same principle. In the description above, when Thomas joined the women on a daylong foraging trip, she further writes that she had been up early to join that group only to discover that they were taking their leisure. She initially did not understand. She had initially thought, if there is work to do, get started on it early. That was not what happened. "So at the Ju/wa camp, we sit around not doing much of anything until almost midmorning when the sun is at forty-two degrees" (Thomas 2006, p. 112). Later she understands: Walk the veld when the predators rest. That meant being active during the hottest part of each day. The Ju/wasi were not at the top of the food chain. During the night, the Ju/wasi took precautions. They slept in a light grass shelter, a *tshu*, which Thomas believes discouraged lions from attacking, for lions like to attack from the side or behind, after having a good view of their prey. The tshu limits the lion's natural inclinations. Also at night, almost always someone in the group was awake, perhaps getting warm by a fire or sipping water from an ostrich eggshell. The presence of fire and the alertness of a person may also have discouraged the lions.

The lions also played a part. They were not particularly interested in eating people. Thomas (2006) offers several reasons. For one, lions preferred larger prey. Not that they would pass up smaller prey if needed, but it was not their first choice. Second, "lions are very intelligent, as are all cats, and also profoundly good observers, and can be open to suggestion, unless they are already excited with their minds made up" (ibid., p. 161). Thus that night when the lions were in the Ju/wa camp, they were at least as curious as they were focused on hunting, and thus could be persuaded to change their plans when challenged with firebrands. Third, and finally, and perhaps most important of all and least often thought about:

Like most other mammals, young lions do as their parents do, preying upon what their parents prey upon, learning techniques and patterns of behavior from their elders. Man-eating is a learned behavior, and if lions of the past didn't start it, their descendants would have a good chance not to be thinking about it and therefore not to pick it up. So if the lions of Nyae Nyae did not hunt people, perhaps it was because their parents hadn't done so. (Ibid., p. 162)

In comparison to the lions of Nyae Nyae in the 1950s, the lions on the African savannahs today are dangerous. The traditional life of the Bushmen is gone; there are no more hunter-gatherers with whom the lions live. The lions lost that cultural knowledge. Then they had to figure out how to deal at best with tourists in game parks, and at worst with people who sought their land and their deaths.

Thomas (2006) emphasizes that cultural guidelines in the animal world can be fragile, "especially those of the Old Way" (p. 167). A final example is worth noting. In the past, there was a bird, the honeyguide, that did pretty much what its name implies. It would locate a beehive, and then locate an animal or person who could help open it. "The bird would fly conspicuously from tree to tree, waiting for the helper to catch up before moving on. At the bee tree, the bird would wait while the honey hunter did the hard work, breaking open the hive, swatting bees, and getting stung. When the melee was over . . . the honeyguide would fly down and eat the larval bees" (ibid.). But now, in current times, people no longer understand what the honeyguides want; and the honeyguides "of today have no memory of the ancient partnership between their kind and ours. . . . So the whole honey-hunting partnership has collapsed and vanished" (ibid.).

That is what I mean by saying that in the Old Way the Ju/wasi cohabited with nature. It means not only coexistence—though that is a profound idea in itself—but that in the same motion people affiliate with the natural world and the natural world affiliates with us.

Conclusion

What is stunning and sad about Thomas's (2006) book *The Old Way* is that a book like hers may never be able to be written again: a book that offers a firsthand account of an indigenous people with a 35,000-year history of living sustainably in a landscape, and who had had virtually no previous contact with Western people. There are no indigenous people left like that in the world; and individuals—like Elizabeth Marshall Thomas—who once had lived with such indigenous people, and known them well, are now all dead, or unfortunately soon will be.

Why should we care about the Old Way? In part, the answer is that our minds and bodies are products of these earlier times, and to an extent largely unrecognized today, the interaction patterns with nature laid down in our ancestral heritage are with us still.

Granted, we as modern people no longer run down a bull eland in 120 degree heat. But we run for fun, we jog, we step onto our treadmills, or we walk. We still hunt. The National Rifle Association is a powerful lobby. We enjoy gardening, and harvesting what the earth brings forth. We forage a bit, perhaps for some huckleberries on a hot summer's day. Some might say we forage at the supermarket. We like to eat meat. We enjoy the seasons turning, the migration of birds. We enjoy the freedom of traveling, moving about. There used to be an airline advertisement on television that ended with the statement: "You're now free to move about the country." That is how the Ju/wasi felt. Though our technology limits the impact of environmental forces, we are still checked by them. Hurricanes can flood our cities; AIDS ravages our people; and children died of starvation today, and yesterday, and will tomorrow. We no longer stare down a lion under the night sky with a burning branch. But vestiges of our desire to be in contact with wild animals, even dangerous predators, remain. We visit zoos for this reason.

Shepard (1998) says: "In a society committed to goals of development and progress, looking back is seen as regressive. Insofar as the past is seen as limiting, the modern temper has never been sympathetic to genetic or essentialist excursions into the complex processes of becoming and being human in the sense of prior biological or psychological constraints" (p. 1). Yet he goes on to argue that both genetic and essentialist accounts "are crucial for the definition ourselves: the human genome is the blueprint that frames our choices of ways of life, of healthy or sick cultures" (pp. 154–155).

We should care about the Old Way because it points not just to our past but—in integration with our technological selves—to our future, or at least the future that we should seek.

2

Biophilia

Still today, in modern times, humans are shaped by the Old Way. That is the basic idea from chapter 1. The idea has been taken up in the academic literature under the term *biophilia*.

This term was used as early as the 1960s by Fromm (1964) in his theory of psychopathology to describe a healthy, normal functioning individual, one who was attracted to life (human and nonhuman) as opposed to death. In the 1980s, Wilson (1984) published a book titled *Biophilia*. I have never seen Wilson cite Fromm's use of the term, so it is unclear whether Wilson was aware of this earlier usage. Either way, Wilson shaped the term from the perspective of an evolutionary biologist. He defined biophilia as an innate human tendency to affiliate with life and lifelike processes. Biophilia, according to Wilson, emerges in our cognition, emotions, art, and ethics, and it unfolds "in the predictable fantasies and responses of individuals from early childhood onward. It cascades into repetitive patterns of culture across most of all societies" (Wilson 1984, p. 85).

In 1992, Kellert and Wilson hosted a small three-day meeting in Woods Hole, Massachusetts, focused on the hypothesis that biophilia exists and could be empirically validated. They brought together about fifteen prominent international scholars, and two young scholars starting out, myself included. That was my first major foray into the ideas of biophilia. Kellert and Wilson (1993) subsequently published an edited volume from that meeting, *The Biophilia Hypothesis*. Much of that material I review in the first two chapters of my 1999 book, *The Human Relationship with Nature*. In this chapter, I review what some of the theory, empirical evidence, and applications look like. Then, from some years of hindsight, I discuss four overarching problems with the term, as a means to understand its potential for the future, and to begin to motivate the empirical methods for my research on technological nature discussed in later chapters.

Theory, Empirical Evidence, and Applications

By most evolutionary accounts, our species emerged hundreds of thousands of years ago on the African savannahs. In turn, certain features of that landscape and ways of being offered greater opportunities for individual and group survival. A person, for example, who could quickly identify and respond to a poisonous snake or the predatory interest of a leopard had greater chances for survival than a person who could not. In this way, biophilia evolved through what Wilson (1993) calls *biocultural evolution*, "during which culture was elaborated under the influence of hereditary learning propensities while the genes prescribing the propensities were spread by natural selection in a cultural context. . . . A certain genotype makes a behavioral response more likely, the response enhances survival and reproductive fitness, the genotype consequently spreads through the population, and the behavioral response grows more frequent" (pp. 32–33). In other words, genes that lead to behaviors that enhance survival tend to reproduce themselves (since they are in bodies that procreate more rather than less), and thus these genes and correlative behaviors grow more frequent.

As evidence, studies have shown that people are innately predisposed to acquire and especially not to forget a fear of dangerous animals, such as snakes. People also respond physiologically to photographs— presented so quickly such that the conscious mind cannot identify the specific images—of fear-inducing animals, such as snakes, but not fear-inducing artifacts, such as guns (see Ulrich 1993 for a review of numerous conditioning and counterconditioning experiments).

Based on preference ratings for different sorts of landscapes, people tend to prefer natural environments over built environments, and built environments with water, trees, and other vegetation over built environments without such features (Kaplan and Kaplan 1989). There is even some evidence "that people have a generalized bias toward savannah-like environments" (Orians and Heerwagen 1992, p. 560). For example, in one study people were asked to rate the attractiveness of different kinds of trees. The researchers found that the trees rated as most attractive—those with moderately dense canopies and trunks that bifurcated near the ground—matched the prototypic savannah tree (Orians and Heerwagen 1992). In another study (Balling and Falk 1982), researchers found that children (8- and 11-year-olds) raised in a nonsavannah setting (the eastern United States with mixed hardwood forests) preferred a savannah biome over their own, as well as over rain forest, boreal forest, and desert. In turn, older children equally preferred the savannah and their own biomes over the other three, supporting the biocultural proposition that habitat preference is shaped by both evolution and experience.

If through evolution certain natural landscapes have promoted human survival and reproductive success, then it would seem likely that such landscapes nurture people physically and psychologically. Research supports this proposition. In one study, for example, patients after gall bladder surgery were assigned to one of two hospital rooms (Ulrich 1984). One room looked out on a brick wall. Another room looked out on a stand of deciduous trees, in a parklike (savannah-like) setting. Based on numerous measures (e.g., self-reports, nurses' reports, amount of pain medication used, and length of the stay in the hospital), patients who had the nature view did better. In another study (Ulrich et al. 1991), people were exposed to a stressful movie and then to videotapes of either nature or urban settings. Based on physiological measures (heart rate, muscle tension, skin conductance, and pulse transit time), results showed greater stress recovery in response to the nature setting. Findings from over 100 studies have shown that stress reduction is one of the key perceived benefits of spending time in a wilderness area, especially in those settings that resemble the savannah (Ulrich 1993).

Many of the stories Thomas (2006) tells of the Ju/wasi living the Old Way involve animals. There are stories of varied hunts of different animals across changing seasons; of predators, such as the hyena, leopard, and lion; and of poisonous snakes, such as a python. There are also small stories of lizards, tortoises, mongooses, rabbits, partridges, and guinea fowls.

Does the biophilic predisposition to affiliate with animals exist in modern people? It would be hard to think otherwise. In the United States, dogs and cats are common pets, and they require substantial time and finances to keep. Supermarkets often devote an entire aisle to pet food. Pets matter to people. In North America, more people visit zoos and aquariums than all sporting events combined. Research shows that contact with animals promotes physical and psychological health for people with Alzheimer's disease, for children with autism, for patients before surgery, and for the general population (Katcher and Wilkins 1993; Melson 2001; Myers 2007). Animal imagery is deeply embedded in modern human language. For example, we use "expressions like porker, hogwash, male chauvinist pig, gas hog, road hog, living high on the hog, happy as a pig in muck, going hog wild, piggish, and crying like a stuck pig. There are fascist pigs and Nazi pigs; prostitutes and policemen are called pigs" (Lawrence 1993, p. 325). That is just for pigs. It works for other animals, as well. Lakoff (1987, pp. 409–411) points to the following expressions of animal imagery: "Don't touch me, you animal!" "He's a wolf." "Stop pawing me!" "Hello, my little chickadee." "She's a tigress in bed." Claude Levi-Strauss says that animals are "good to think" as well as good to eat. That dictum holds true not only for primitive cultures but for complex modern societies.

Currently, one interesting application for biophilia uses the theory and empirical base to structure the design of the built environment. Joye (2007), for example, writes of the need for "biophilic architecture" (p. 324), which can be achieved in two ways: One, bring aspects of nature into the built environment (e.g., flowers, trees, and water); second, ferret out the key structural properties of nature (e.g., in terms of prospect and refuge, mystery, legibility, and fractal patterning) and mimic those properties when we build. Along slightly different lines, Kellert (2005) has offered a vision of "biophilic design" that includes what he calls *organic design* and *vernacular design*: "*Organic design* involves the use of shapes and forms in buildings and landscapes that directly, indirectly, or symbolically elicit people's inherent affinity for the natural environment" (p. 5). This effect can be achieved, for example, through the use of natural lighting and the presence of water and vegetation. "*Vernacular design* refers to buildings and landscapes that foster an attachment to place by connecting culture, history, and ecology within a geographic context" (ibid.). These ideas have been extended in Kellert, Heerwagen, and Mador's (2008) edited volume entitled *Biophilic Design*.

From the vantage point of several decades of work on biophilia, I would like to offer where I see this body of work: its conundrums and its future.

Four Overarching Problems

As a unifying construct, biophilia has been successful. It has allowed research from many different disciplines on the human relationship with nature to be brought together, seen from a larger vantage point, and sometimes integrated. Biophilia has also provided a theoretical grounding to the empirical research. It has also been leading to important applications, especially in terms of the built environment.

Nonetheless, there are some problems with the construct. In chapter 2 of my 1999 book, *The Human Relationship with Nature*, I described nine potential conceptual difficulties and empirical limitations of biophilia; for most of them I showed how the problems are not nearly as large as critics suppose. But now, given more years of "biophilia under my belt," I think the construct has four overarching problems.

The "Bio" of Biophilia

The first problem is of how to understand the prefix "bio" in biophilia. Wilson (1993) is clear that he means biological life. In his chapter "Biophilia and the Conservation Ethic," for example, Wilson states that biophilia "is the innately emotional affiliation of human beings to other living organisms" (p. 31). In *Naturalist*, Wilson (1994) states that biophilia "means the inborn affinity human beings have for other forms of life" (p. 360). Wilson is a biologist. It is not surprising that he focuses on life. But people affiliate with nonliving nature, too: with caves, canyons, mountains, and geysers; with rocks, sand, wind, steaming gorges, hot springs, mud pits, glaciers, tundra, and salt flats (Ruckert and Kahn 2007). People explore and study these nonliving natural phenomena. People bond with them. People work to conserve them. Consider the world's largest monolith, Uluru, the site of an aboriginal creation story and one of the world's major tourist attractions. In Ruckert's telling, Australia is partly defined by this large red rock. What is it that draws people to it in the middle of nowhere? Why do they want to touch it, to climb it, to see the sun rise behind it, and even to worship it? One does not ask these questions if "biophilia" is taken literally to mean an affiliation with life.

There is an easy fix, which both works and does not work. The fix is that we simply need to redefine biophilia to include nonliving nature. Most proponents of biophilia probably make this move anyway, and hardly think about. But it does not work insofar as it undermines the elegant clarity of the term itself, biophilia, which specifies living organisms. For accuracy, the term would need to be renamed. But to what? Naturephilia? Envirophilia? Ecophilia? Nothing seems as clean for a prefix as the "bio" in biophilia.

The "Philia" of Biophilia—Does a Construct of Biophobia Need to Be Invoked?

Here is another problem. The suffix "philia" means having a tendency toward, to like, and to love. I suspect most people think of "philia" in biophilia in a positive, upbeat way. Yet clearly there are many aspects of nature that we think of in a negative way. We might dislike being cold in the rain, being threatened by lightening, or breathing the smoke of a nearby forest fire. Few of us enjoy mosquitoes. The thorns of a rose bush may bring us no pleasure. We would all fear, just like the Ju/wasi, the charge of a lion. For this reason, some proponents of biophilia have sought to distinguish biophilia from what they call *biophobia*. Orr (1993), for example, says that biophobia "ranges from discomfort in 'natural' places to active scorn for whatever is not manmade" (p. 416). Ulrich (1993) speaks of "adaptive biophobic (fear/avoidance) responses to certain natural stimuli and situations" (p. 85).

Wilson himself has been unclear on what he means by "philia." Sometimes he subsumes both positive and negative interactions with nature within the construct of biophilia. For example, in his chapter "Biophilia and the Conservation Ethic," Wilson (1993) writes that biophilic feelings "fall along several emotional spectra: from attraction to aversion, from awe to indifference, from peacefulness to fear-driven anxiety" (p. 31). Similarly, in *Naturalist*, Wilson (1994) includes in the definition of biophilia the phrase "fascination blended with revulsion" (p. 362), and he points to our relationship with snakes as having both qualities. Yet other times Wilson refers to the positive aspects of human–nature interaction as biophilia and the negative aspects as biophobia. For example, in *The Future of Life*, Wilson (2002) writes:

The companion of biophilia is therefore biophobia. Like the responses of biophilia, those of biophobia are acquired by prepared learning. They vary in intensity among individuals according to heredity and experience. At one end of the scale are mild distaste and feelings of apprehension. At the other end are full-blown clinical phobias that fire the sympathetic nervous system and produce panic, nausea, and cold sweat. (p. 141)

Separating the two terms—biophilia and biophobia—seems so obviously sensible. After all, some nature interactions we like and other nature interactions we do not like, and the two terms seem to say just that.

But our minds and bodies came of age in response to many biophobic stimuli, and we need them still today for our well-being. An analogy can be made to exercise. I suggested earlier that one of the defining characteristics of the Ju/wa people's interactions with nature is that they used their bodies vigorously. In modern times, we still have bodies that thrive better by means of an active lifestyle—biking, swimming, running, or going to the gym and lifting weights or getting on the stairmaster. Exercise is life-affirming even if some people find it unpleasant. The same case can be made for many biophobic interactions, though I am not aware of any research that has yet explored this hypothesis. But think of it: Jumping in a cold river is initially unpleasant, and then often invigorating. Walking under the night sky can be a little unnerving insofar as we lose much of our sense of sight, but it can also be awe-inspiring. Walking a land where rattlesnakes are present makes the mind not stressed, but more alert, calmly focused—at least I have found that to be the case. Thomas (2006) writes that the Ju/wasi were aware, at times concerned, but rarely stressed by living in the presence of dangerous animals. Indeed four of the six aspects of Ju/wa life that I characterized earlier—using bodies vigorously, periodicity in the satisfaction of physical needs and of nature, environmental checks and balances, and encountering the wild—all have aspects that can be experienced as unpleasant, as "biophobic." Yet all of them can also be thought of as biophilic. Rolston (1989) puts it this way: "Environmental life, including human life, is nursed in struggle; and to me it is increasingly inconceivable that it could, or should, be otherwise. If nature is good, it must be both an assisting and a resisting reality" (p. 50).

That is the problem of separating biophilia from biophobia. It cleaves apart what is essentially a unified experience. For this reason, I think a more productive, accurate, and elegant approach is to have "biophilia" refer to both positive and negative interactions with nature. Jettison the term "biophobia." Then take up the challenge of how to understand the coexistence and integration of positive and negative interactions with nature within an orientation that is life-enhancing.

Can Biophilia Ever Be Disconfirmed?

The short answer is: probably not. Imagine we collected evidence that people had a tendency to affiliate with their cars, or with music, or with members of the opposite sex. Would that evidence counter the biophilia hypothesis? Not at all. To say that people affiliate with nature is not to say that people affiliate only with nature. Thus, as long as nature effects can be found in enough relevant situations—and they can be—biophilia cannot lose. But in this way, biophilia is not a scientific hypothesis, at least not in a stringent sense of being predictive, testable, and open to disconfirming evidence.

That is troubling to some (Fischer 1994). It used to be troubling to me. But I now think that biophilia is best understood not in itself as a testable hypothesis (no more so than the idea that people have an affinity for other people), but as a broad construct that helps to generate hundreds of important testable scientific hypotheses. That is, the science happens at a lower level. For example, Ulrich (1984) hypothesized that people would heal faster when recovering from surgery in a hospital room with a view of a savannah-like park out the window compared to a brick wall. That was a tractable hypothesis. Thus, we may need to move away from the term *the biophilia hypothesis*—a term that Kellert and Wilson (1993) worked hard to put on the map.

Biophilia and Genetic Determinism

If one looks at the theoretical underpinnings of biophilia, as Wilson conceives of it, one could be led to believe that not just biophilia but all behavior is genetically determined, and that the mind operates mechanistically.

According to Wilson (1993), biophilia evolved in the human species by means noted above: genes that led to behaviors that enhanced survival tended to reproduce themselves (since they were in bodies that procreated more rather than less), and thus those genes and correlative behaviors grew more frequent. That explanation lies at the core of sociobiology, a field Wilson (1975) helped create. Dawkins (1976) phrases the idea this way: "We are survival machines—robot vehicles blindly programmed to preserve the selfish molecules known as genes" (p. ix). According to this view, we as individuals do not have choice, autonomy, or free will. True, sociobiological theorists recognize that people use these terms all the time, and these theorists recognize that many people believe they have choice, autonomy, and free will. But these beliefs, the theorists argue, are epiphenomenal, meaning that they play no authentic causal role in human action. To stress this point, Dawkins compares a person to a guided missile, which appears to search actively for its target, "taking account of its evasive twists and turns, and sometimes even 'predicting' or 'anticipating' them. . . . Nothing remotely approaching consciousness needs to be postulated, even though a layman, watching its apparently deliberate and purposeful behavior, finds it hard to believe that the missile is not under the direct control of a human pilot" (p. 54). Even behaviors that seem altruistic are, according to sociobiologists, selfish, and have been put into place by means of the evolutionary adaptive process. As Ruse and Wilson (1985) write: "Morality, or more strictly our belief in morality, is merely an adaptation put in place to further our reproductive ends. . . . In an important sense, ethics as we understand it is an illusion fobbed off on us by our genes to get us to cooperate" (pp. 51–52). In the conclusion to his treatise on sociobiology, Wilson (1975) says that in the final analysis psychology, sociology, and the other human sciences will be reducible to neurobiological processes: "Cognition will be translated into circuitry. Learning and creativeness will be defined as the alteration of specific portions of the cognitive machinery. . . . To maintain the species indefinitely we are compelled to drive toward total knowledge, right down to the levels of the neuron and gene. . . . [When] we have progressed enough to explain ourselves in these mechanistic terms, the result—'a world divested of illusions'— might be hard to accept, but true" (p. 575).

Wilson, as he is wont, was prescient. Now, many decades after he wrote those words, it could be said that the core fields of which he speaks are being reshaped, redefined, and usurped in ways that he foresaw.

The problem is that I think Wilson and others are wrong about how best to understand and investigate the human mind. To convey my reasons, I provide a short debate between myself (PK) and a genetic determinist (GD). Though this particular debate is imaginary, it follows the many discussions I have had with colleagues who have professed some version of biological determinism.

PK: You say we're gigantic lumbering robots, controlled by our genes, and that we have no more choice in our behavior than does a guided missile. But here is a simple example of a choice, which I'm modifying and extending from one offered by the philosopher John Searle in one of his lectures. Let's say that I'm thinking of the alphabetical letter *Q* in my head. I want to type this letter, using my computer. I have the intention to type it. There is no one telling me to type it. If I type it, it will be my choice to type it, mine alone. Here goes: *Q*. I did it. I typed it. I had a choice to type *Q* or not. I chose to type *Q*. So, sure, language may well have an innate structure, and its genesis lies in our evolutionary history. But it makes little sense to say that I have an innate genetic predisposition to type the letter *Q*, or that my genes determined this act. I could just as easily have decided not to type the letter. Indeed, I can do that right now. I am *not* going to type that letter. There, I did not type it. I chose not to. I am not like a guided missile.

GD: That's a pretty trivial example. That's not the level at which most of us genetic determinists are talking about choice.

PK: But once you grant it, you're in the mix. I choose whether or not I type the letter *Q*. I also choose whether or not I give money to a homeless person on the street. Sure, there are innate aspects of altruism; and I can see your point that in ancestral times some altruistic behaviors may have enhanced genetic fitness. But here and now, I still have a choice. I can give money or not. I can also talk to other people and they can talk with me. We can make choices together. Three of us, for example, could talk things over and choose to go to a movie, or to volunteer in a homeless shelter. We could as a local community choose to make recycling easier. We could as a world community decide to save endangered species. My point is that we make choices about trivial and nontrivial matters. We are in bodies and minds that are capable of making choices.

GD: Hey, I understand it feels to you like you're making a choice, and that you have that sort of freedom. That's a powerful feeling, of course. We all have those feelings. But what I'm saying is that in ancestral times it enhanced survival and reproductive fitness to believe deeply that we had this sort of free will; thus the genes prescribing the propensity for such beliefs were spread by natural selection. Just believing something doesn't make it true. Some people believe that aliens have visited the earth in flying saucers. Our beliefs can be wrong. Can you prove that you have free will?

PK: No, I can't knockout prove it. But there is good reason for us to believe in such internal states that are concordant with our lived verifiable experience. That's different than aliens visiting in flying saucers. Can you prove that right now you're not a brain in a vat being stimulated by electrical currents to believe that you have a body and are reading these pages? No, there's no knockout proof here, either. But at the very deepest level of our being we believe we exist, and there's a huge amount of evidence that supports our beliefs: it is the entire history of our lived lives. Why should we discount a lifetime of evidence in lieu of a fanciful proposition about a brain in a vat? Ditto with choice, autonomy, and free will. Also, you just argued that in ancestral times it enhanced survival and reproductive fitness to believe falsely that one had free will. That's a little odd, don't you think? I mean, why would such beliefs in choice, autonomy, and free will ever have been adaptive in ancestral times? It could just as easily be said that such beliefs unduly complicate the mental apparatus that is better served addressing more immediate problems of survival. It's here that you're offering a "just so" story: a post hoc account without enough specificity to rule out competing and sometimes more compelling explanations.

GD: I'm surprised by your distrust of these "just so" stories. You tell them all the time. In your writing on the Old Way and biophilia, you often say that our propensities to affiliate with nature—for example, that we're drawn to bodies of water, flowers, and places of prospect and refuge—are with us today because they were adaptive in ancestral times. What's the difference between what you do and what I do?

PK: The difference? You're reducing humans to a programmed biological machine. I'm not. But, having said that, I am also committed to the proposition that we're biological beings with an evolutionary history that still today shapes and constrains and offers particular affordances for human life. So, yes, I tell some "just so" stories because some of these stories seem compelling, and may well be true. I would add that the benefit of paying close attention to Thomas's (2006) account of the Old Way, and others like it, is that it helps minimize our post hoc accounts insofar as we see evidence from the 1950s of an actual people (such as the Ju/wasi) that were living sustainably in ways that may have existed over 30,000 years ago.

GD: We agree about the importance of understanding the Old Way. But on this machine business, we genetic determinists aren't as bad as you think. Our "machinery" actually encompasses the human spirit. We're humanists, too. For example, Wilson (1992) writes that an "enduring environmental ethic will aim to preserve not only the health and freedom of our species, but access to the world in which the human spirit was born" (p. 351). Before that he writes that we "do not understand ourselves yet and descend farther from heaven's air if we forget how much the natural world means to us. Signals abound that the loss of life's diversity endangers not just the body but the spirit" (p. 351). Elsewhere, Wilson (1984) writes that "our spirit is woven from it [biophilia], hope rises on its currents" (p. 1). Freedom, spirit, heaven's air, and hope rising on biophilic currents. Aren't you taken with the beauty of his words?

PK: Sure, they're beautiful words. But here's my problem: Wilson charms people with them. People read these words and think "hey, biophilia speaks to me, it's so beautiful, my spirit soars, I support biophilia." But Wilson must know that people interpret his words in ways that he does not mean. Whatever Wilson means by "spirit"—and I'm not sure he exactly knows himself—it is bounded by a conception that nothing human exists beyond the biological body; there is no transcendent self, spirit, or soul; and there is no other world of heaven. But to be clear,

I am not addressing here the issue of whether there is a transcendent self or reality (e.g., of whether morality exists independent of people or whether a god exists). Remember, I'm focused on the basic issue of whether human nature encompasses choice, autonomy, and free will.

GD: I remember. But you need to understand that you don't exactly understand our work. First, part of what we're doing is modeling the human mind and the evolutionary process. Modeling is a lot of what science is about. You're a scientist. Why are you taking issue with us here? Second, while some of us still speak of genetic determinism, most of us understand our ideas in terms of biocultural evolution, where there is an interaction between genes and culture. So you're mischaracterizing us. And, third, even if we imply genetic determinism, it's quite a minor part of what our work is about.

PK: There's a lot to respond to. I'll be as brief as I can. First, models can be great. And perhaps you're right that at times it could be useful to think of ourselves *as if* we were programmed biological machines. But there should be no confusion between the model and the mind. As Searle (1990) notes, it can be useful to model water molecules with ping-pong balls in a bathtub, but you'd be silly to expect to get wet stepping into a tub of ping-pong balls. Second, while many of you do speak in terms of biocultural evolution, the real work is usually still happening on the level of evolution and genetic fitness, and there's a lot of hand-waving about culture. Also, in your conception of culture—for example, that cultures transmit knowledge or condition behavior—the individual is still considered a passive entity, without choice, autonomy, or free will. So even if you put forward a robust version of biocultural evolution, the individual is still controlled, only now by some interaction of internal forces, the genes, and external forces, culture. Third, small is big. Here's an analogy. Biologists say that 99 percent of our DNA code is the same as that's in a chimp. If we were just playing percentages, we could say there's virtually no difference between our two species. But I'd say that 1 percent is a huge difference. It's a difference of a nuclear bomb, Shakespeare, genocide, sprawling cities, iPhones, Van Gogh, General Motors, and brain surgery. I agree that we are deeply shaped by our evolutionary past, and constrained in certain ways by it. But even if there is only a small part of us that is "free"—it's an absolutely crucial part of our human nature. We need to affirm it, and exert it, as a means to solve our global problems and to flourish as a species.

GD: So, if you believe all these things, then what's your solution?

PK: I appreciate your asking. I'll take up that question in chapters 11 and 12.

Conclusion

Without doubt, the term "biophilia" has advanced the field. But given the four overarching problems above with the term, does biophilia have enough left in it for another good run, for say another twenty years of usage? I am not sure.

One option is to coin another term. For example, there is currently a large field focused on people's interactions with computational systems. It is called *human–computer interaction*. There is an emerging field focused on people's interactions with robots. It is called *human–robot interaction*. Thus perhaps it would be worthwhile to substitute for "biophilia" the term *human–nature interaction*. By being more descriptive, "human–nature interaction" sidesteps the problems of "The 'Bio' of Biophilia" and "The 'Philia' of Biophilia" described above. But the problems of the term "biophilia" also partly constitute its strength. "Biophilia" is a "feel good" word about our relationship with nature. People can easily grab hold of the term, easily speak about it, and feel like they know something about it. People can advance its cause. Thus another option is that we keep using the term "biophilia," while knowing that it does not accurately capture what we mean, though it captures some of it. We do that with other expressions. We speak for example of a *sunrise* because the term captures our visual experience, though in terms of the physics of the situation it would be more accurate to say an *earth turn*.

Whether the term "biophilia" gives way to another term or not, the underlying ideas I believe are immensely important. Our minds and bodies came of age hundreds of thousands of years ago, and thrived through patterns of interactions with the natural world. We cannot jettison all of these interaction patterns and still thrive as a species. Some of these interaction patterns were comforting, others invigorating or harsh or soothing or satisfying or fear-inducing, and some were all of these experiences combined. I will say more about these interaction patterns in chapter 12, where I propose the framework of a *nature language*: a systematic way of speaking about human–nature interaction. In this nature language, and other approaches as well, it will be important to build on the success that proponents of biophilia have had in unifying diverse fields of research, grounded in evolutionary theory. It will also be important to integrate a nondeterministic and nonmechanistic conception of human beings. We make choices. We will create our future.

3

The Technological Turn

At some junction in our evolutionary history, we began to create, use, and love technology. That is what I mean by the *technological turn*. It is odd, though, for while we are a technological species, our technologies do not always appear to advance our lives.

Consider this example from about a hundred years ago (Ingalls and Perez 2003; "The Tradition of the Lector," 2004). When the Cuban cigar factories of the late nineteenth century made their way to Florida, they carried over the unique tradition of the "lector." This individual was hired (and well paid) by the workers to read aloud during the day. Workers chose the readings. Typically, morning readings comprised political tracts, and afternoon readings comprised a classical repertoire, including Shakespeare, Tolstoy, and Cervantes. Each reading was in effect a performance. The lector would feel the tenor of the audience, and modify his delivery, extend the reading or cut it short, for maximum effect. Listening to the lector helped ease the monotony of the work, day in and day out. Through the lector, the workers, who were often illiterate, became well educated in political events and could often quote from the classical texts. The readings also led to interesting conversations at the day's end, all of which they could then bring back to their families by evening.

The factories had little technology. The cigars were made by hand. The factories were small enough so that a single lector could be heard easily by all.

It is a common human problem, of course: how to stay engaged and alive to the world when confronted with a job that is mundane and repetitious. Those in the cigar factories had found a solution that worked for them.

But these factories began to face competition. With the expansion of industrialization, parts of the cigar-making process could be done by machines, which at a large enough scale brought prices down. To survive, other factories had to mechanize and become larger. Because the physical size of the factories increased, and because the machines made noise, it became difficult to hear the lector. The lector disappeared. The modern assembly line emerged.

The term *robota*—the origin of the term *robot*—was coined by the Czech playwright Karel Čapek in the 1920s to refer to the mechanized dehumanization of industrial labor. Some have said that our technology, meant to reduce human drudgery, became our masters, and we in effect became the robots. Some look on such uses of technology and see corporate greed as a driving mechanism. Others see a populace that demands cheaper and cheaper consumer goods, without care as to their origins.

By "technology" I mean any human artifact that assists humans to achieve an effect in the world. Technologies can be very simple. The Ju/wasi digging stick, for example, is about three feet long, an inch in diameter, and pointed at one end. Thomas (2006) says that it is better than a shovel for digging roots in the African savannah because it is light and you can feel what you are doing in the earth. A hammer is also a technology, as is a paper clip, spatula, air conditioner, pencil, gun, corkscrew, chainsaw, lightbulb, stick of dynamite, and motorboat. Not all objects are technological. A rock sitting by the river is not a technology, for it is natural, not a human artifact. A tomato is not a technology, for it is biological, not artifactual. Even a spool of metal wire by itself is not a technology, for it has not yet an implied human function; though when we use it to snare a rabbit, it becomes a technology. Of course, many of our technologies today are more complex than digging sticks and metal snares. Rather, they include computation and microelectronics, and give rise to computers, cell phones, Wi-Fi, satellites, search engines, software agents, telesurgery, robot warriors, and thought-controlled prosthetics. Computation and microelectronics are also increasingly embedded in earlier forms of technologies, from ovens to cars to artificial hearts. In my definition, I also include information technologies. They, too, can be simple: a smoke signal, a cave painting, a book; and complex, as with telecommunications; and embedded, as with genetic engineering.

We cannot survive without technology. Indeed, as the decades unfold, we will not be able to survive without increasingly pervasive, complex, and computational forms of technology. But the technological turn also brought with it the problem I mention above: we create technologies that often hurt us. It would be good if it could be otherwise. I think it can be. Toward that end, this chapter aims to provide enough intellectual background and practical methods on the technology-side of "technological nature" so as to situate my empirical studies of chapters 5 through 10 and lead to my discussions at the end of this book on how we can flourish as a technological species in relation with nature.

Why Did the Technological Turn Happen?

I begin by asking why the technological turn happened, because by knowing our origins we learn something of ourselves as we are today.

Even if our historical records were complete, it would be somewhat arbitrary to say when the first technology emerged. If we wanted to stretch very far, perhaps we would go back about 2.5 million years, when most "paleontologists believe that simple chipped-stone pieces associated with the remains of *Homo habilis* indicate the dawn of tool-making" (Ehrlich and Ehrlich 2008, p. 66). About 1.6 million years ago, *Homo erectus* is believed to have first controlled fire. About 50,000 years ago, according to Ehrlich and Ehrlich, *Homo sapiens* deliberately used bone, ivory, and shell objects to shape projectile points, needles, and awls, and engaged in cave painting and sculpture. More recently, we could point to the tools used to domesticate land and animals during the Mesolithic period about 15,000 years ago. By about the middle of the third millennium BCE, there were small two story houses, and shortly thereafter the Chinese used blast furnaces to cast iron. By the sixth century CE we had the iron plow, and by the thirteenth century the spinning wheel. The Western Renaissance emerged in the 1700s, and after that was the industrial revolution. That gets us to the cigar factories in Florida in the 1920s. Perhaps the greatest amount of technological innovation in the shortest period of time has occurred in the last fifty years, or even in the last twenty years. Regardless of how we chart it, technological innovation remained reasonably stable for an extraordinarily long period in our evolutionary history, and then began rapidly to escalate, increasingly so, in its sophistication and pervasiveness. Why?

We could try to answer this question the way we did for biophilia in chapter 2: the behavior was adaptive. After all, it could be said that a predisposition for technological innovation conferred genetic advantage on a people. For example, if you could invent a better tool by which to hunt, or to farm, or by which to control animals, such as an elephant or horse, then that invention would increase the chances of survival for yourself, your children, and your group, at the expense of other people and groups less inventive. In turn, without any conscious intent, those genes that promoted such innovation would be passed along to the next generation. By this means, we emerged as a technologically innovative species.

But if this explanation is correct, then why were many indigenous people, at least until recent years, largely nontechnological? Why did the Native Americans not invent rifles for hunting? Why did the Inuit in Alaska not invent snowmobiles for travel and hot tubs for relaxing? Thomas (2006) answers this question by inverting the adaptive account. She writes: "Although our bodies changed as we became human beings, and although we changed many of the things we thought and did, we didn't change anything unless we had to, because change for its own sake is undesirable, experiments are risky, and life is tenuous enough without departing from what is known to be helpful and safe" (p. 10). Put as a syllogism, her account goes something like this: If a current lifestyle works, don't change it. The indigenous lifestyle worked. Therefore, it didn't change.

The puzzle is as follows: If it was not more adaptive to innovate technologically, as Thomas suggests, then why did most of our species technologically innovate? If it was more adaptive to innovate technologically, then why did some people (like the Ju/wasi) not innovate, or (like the Inuit) innovate only minimally, and why did the technological turn not happen much earlier in our evolutionary history?

I would like to offer the sketch of an answer, which builds on others, though with a few new lines. The answer is speculative. I do not see how it could be otherwise, for the historical data are too scant.

I begin by noting Thomas's (2006) account of the social fabric of the Ju/wasi. Most of us are members of many different groups—with family and friends, and in our jobs and elsewhere—and if we fall out of favor with one group, we can fall back on another, or create a new group, or at least still survive without any group, albeit more lonely than we might like. But the Ju/wasi had only a few groups to choose from, and their survival depended on high levels of cooperation. Thus, according to Thomas, the Ju/wasi depended on one another emotionally and physically in ways that far exceed our own experience. Moreover, as a means to maintain group harmony, equality permeated Ju/wa social etiquette and practice. A hunter, for example, did not own the meat he killed. "That role belonged to the person who provided the arrow that actually killed the [animal]. . . . By the Ju/wa system, anyone could own an arrow or arrows (although only the hunters used them), so that an old man or a woman or a boy like Lame ≠Gao . . . who had little chance of ever being much of a hunter, could give an arrow to a hunter and become the distributor of important meat" (p. 101). Of course, Ju/wa men and women had different abilities and talents; but always they sought to downplay any expertise, not wanting to stand out from the group, not wanting to engender jealously or conflict.

When *Homo sapiens* moved beyond the African savannahs around 60,000 years ago, their natural resources increased. That by itself encouraged a modicum of technological innovation, for the downside of any particular failure was not fatal, and the upside large. In addition, the increase in natural resources allowed people to stay located in a single spot. That is, we shifted from being nomadic to domestic. In turn, with greater natural resources and a domestic lifestyle (such that we did not have to carry our children long distances), it was possible to have larger families. Populations grew.

At that junction, Mumford (1961) suggests that two paths—of the village and of the citadel—were then open for the development of human culture. In my view, both paths promoted technological innovation, though the latter more so than the former. The path of the village comprised "voluntary co-operation, mutual commendation, wider communication and understanding: its outcome would be an organic association, of a more complex nature, on a higher level than that offered by the village community and its nearby lands" (p. 89). In contrast, the path of the citadel "was that of predatory domination, leading to heartless exploitation and eventually to parasitic enfeeblement: the way of expansion, with its violence, its conflicts, its anxieties, turning the city itself into an instrument, as Childe properly observes, for the 'extraction and concentration of the surplus'" (p. 89). Even Athens, for example—a seeming bastion of democracy—is believed to have had, at its height, only 40,000 full-fledged citizens (male), compared to perhaps 150,000 free people without full privileges of citizenship (e.g., women and children) and 100,000 slaves. According to Mumford, this second path has largely dominated the last 15,000 years of "urban history till our own age, and it accounts in no small degree for the enclosure and collapse of one civilization after another" (p. 89). That said, Mumford also extols cities for the very surplus and specialization that they accord, and for the huge intermixing of people and ideas. Cities thus created conditions for enormous creativity and technological innovation.

In summary, our species came of age with scarce natural resources on the African savannah. In that environment, Thomas is correct that change for its own sake is highly risky and is to be avoided. But some modicum of technological innovation became adaptive when natural resources increased. Moreover, the increase in natural resources allowed people to shift from a nomadic to a domestic lifestyle; and both together allowed populations to grow, which then required new social-organizational structures. The path often chosen was that of the city, which was at once exploitive, competitive, and creative. In the city, more so than in the village, technological innovation flourished, and it still does today.

Something Gained, Something Lost

Technologies have been and often are a wonderful boon to human life. Technologies allow us to cut trees, transport them, mill them, and then use the timber for our dwellings, to protect us from the wind, rain, and cold. Telephones connect us to our distant loved ones. Eyeglasses afford us clear vision. Refrigerators keep our food from spoiling. Dental technologies keep our teeth intact. Espresso machines make for tasty brew. Technologies allow us to write books, with ink or by computer; and to read books, clothbound or digital. Technologies keep our babies from dying. This list could go on for many pages. Perhaps my point is obvious to the reader, but I say it anyway, to be clear that I am for technology, which needs to be emphasized because I am also keenly aware that with almost all technologies we not only gain but lose.

I love books. I think of them as almost holy, and it is almost miraculous that the complete lifetime writings of a genius, like Shakespeare, can be made available to anyone for a mere handful of dollars. The printing press made that possible. There is genius, too, in the *Iliad* and the *Odyssey*. But these narratives had originally been spoken, not written. That seems almost like a miracle, too: a mind that could remember an entire story, pretty much word for word, and tell it straight out. As technology allowed the printed word to take hold, we lost the spoken narrative, the oral tradition. We lost that capability. We lost that special pleasure.

Electric lights illuminate the dark. I love the feeling of being inside my cabin with lights ablaze, writing at my desk far into the night. But those same lights blind me to the night sky. And when millions of us in the city keep our lights aglow, the stars can no longer be seen. A friend of mine mentioned to her friend about seeing a shooting star. Her friend truly thought she was kidding, and said that shooting stars were imaginary fancies, like unicorns. If light pollution covers the globe, perhaps many of us will come to think so, too.

In terms of our relationship with nature, what technology perhaps most strikingly diminishes is the richness and variety of the human–nature interaction patterns described in chapter 1. I would like to return to them just briefly in this context.

Using Bodies, Vigorously
Technologies make our lives physically easier. That seems good. It is easier, for example, to get warm on a chilly day by turning up a thermostat on a gas furnace than by chopping wood. Yet in our technological world many people are woefully out of shape, which often diminishes their physical health and psychological well-being.

Periodicity in the Satisfaction of Physical Needs and of Nature
Strawberries in winter? Not a problem. They are grown in another hemisphere and flown to our local market. Technologies allow many of our desires to be fulfilled daily, which might seem good at any moment,

but which can diminish the same experience less often encountered in the context of natural rhythms.

Freedom of Movement

It might seem that technologies fundamentally make freedom of movement easier than in the Old Way: get in a car, boat, or plane, and off we go across land and water, or through the air. But these technologies in part compensate for the geographical rigidity of our daily lives. Domestic life roots us in place. In the Old Way, conflicts were often avoided by moving. In modern times, conflicts are often avoided by walling ourselves in. Walls partition us from one another in our houses, condominiums, and apartments. Even our technological means of moving about is not as free as it might appear, and not just monetarily, as all of us know who have spent time in congested traffic and airports.

Environmental Checks and Balances

I described in chapter 1 how the Ju/wasi occasionally had to practice the act of //kao, of putting down a baby at birth, killing it, if the baby was born while the mother was nursing another. It was not possible for a Ju/wa mother to nourish both young children, nor to live a nomadic life carrying two young children. As I note above, greater natural resources allowed mothers to care for more than one young child at a time. That is a blessing for life. But now through our technological advances we have over six billion people on this planet, which may at best dampen the human experience of life, and may at worst be the cause for tremendous human suffering and death.

Encountering the Wild

Explorers have often faced a poignant irony toward the end of their lives. With great spirit and verve, they open up new territory. People then inhabit what had been before wild, and perhaps make the area a town or even a city. Then the explorers return, either heroes or forgotten, it does not matter. Perhaps the explorers now see a house beside a spring they had once discovered, or roads alongside a wild river they had once crossed. Perhaps the explorers see meadows where they had once grazed their horses, now paved over with asphalt, or factories belching smoke on land that had once been old growth, now clear cut. The sad realization, which I imagine can hit them hard, is that their life efforts have led to the destruction of what they love—that which is vast, free, self-organizing, uncharted, not under human control. It is not how they expected to end their lives. The thing with our technologies is that they become more efficacious, while nature stays the way it has largely been. Thus it takes less and less time, and fewer and fewer people, to make inroads into the wild, and to destroy it.

Cohabiting with Nature

We do not coexist freely with that which we control, nor feel in intimate relation with it, nor it with you. That accounts for one of the contradictions of zoos, for example. They have sought to redefine themselves as conservation centers, and among their responsibilities is that they "strive to help society achieve a more sustainable and harmonious relationship with nature . . . [by] inspiring others to celebrate and conserve nature" (Rabb 2004, p. 237). But through a rigid technological infrastructure, zoos keep animals captive and reify the domination of our species over others. Technology can do that.

What I am saying is that technology is a wonderful part of our essential self; but technology can also diminish the depth and richness of human life, and more specifically our relationship with nature.

Here a reader might think me too hasty, and respond: "Wait a moment! We can technologically fix the problems that emerge through our technology."

We certainly try. Yet the fixes often lead to new sets of problems.

Consider again, as an example, the early twentieth-century cigar factories in Florida, and the loss of the lector. Let us imagine at the time that we had a technological wand that implemented our technological fixes. What might we have done? As factories got larger and noisier through mechanization, we might have created a new technology that amplified the lector's voice, and conducted it throughout the factory. Then, to establish cost efficiencies we might have realized that we needed only one lector for hundreds of factories, because our new technology—we might have called it a radio—was accessible to all. We then might have put into our factory many "speakers"—technological, not human—and tuned in to the new entity we just created, a radio station. Have we solved the problem? Partly. But notice that with the lector as disc jockey, our workers have lost the experience of a live performance, where the lector performed in relation with and in response to an audience that could be sensed. Then, too, our radio stations may become controlled by large corporations, often aligned with the same interests as the owners of the factories, such that the lector's political discourse is jettisoned, as is the classical repertoire insofar as "wisdom" for the masses is not considered a productive goal for the working class. Moreover, since the lector's radio program is transmitted to many different factories in many different regions, we might decide to make the content reasonably generic, in effect standardized and simplified. How fictitious is our scenario? Not completely. In 1931, for example, cigar manufacturers in Florida "abolished the lectors, and radios took their place on the factory floor" ("The Tradition of the Lector," 2004, p. 6).

By What Process Does Technology Affect Us?

I ask this question because the answer will shape the means we employ to address the harms that our technologies cause us.

One answer, which I think is mistaken, posits that designers inscribe their own intentions and values into the technology; and once developed and deployed, the resulting technology determines specific kinds of human behavior. This position is sometimes referred to as *technological determinism* (Smith and Marx 1994). To illustrate the idea, consider Latour's (1992) description of the "Berliner lock and key": a device designed to deal with people who forget to lock the door behind them. The device initially works as a normal lock. You put your key into the lock, turn the key, and the door unlocks. But then to remove the key it is necessary to push it through the keyhole to the other side. Then, by moving to the other side of the door, you rotate the key one more turn, thereby relocking the door. Thus, when used, this technology "determines" that people will lock their doors; and it is believed that the very meaning and intentions that designers and builders bring to their task literally become a part of the technology (cf. Appadurai 1988; Cole 1991). To my mind, this answer makes only partial sense. Granted, there are a few technologies, as with the Berliner lock, that are difficult to use for any purpose other than that for which they were designed. But it makes no sense to impute mental states to inanimate human artifacts that only function when acted upon by human forces.

Rather, I would like to advance an interactional account of the process by which technology affects us (Friedman and Kahn 2008). This account has five basic tenets. First, the features or properties that we design into our technologies more readily support certain behaviors and human values while they hinder others. For example, as system designers we can make the choice to try to construct a technological infrastructure that disabled people can access. If we do not make this choice, then we single-handedly prevent such behavior and undermine the human value of universal access. Second, the technology's actual use depends on the goals of the people interacting with it. For example, most of us most of the time use a screwdriver as a tool to screw in screws. That is what the tool was designed to do. But some of us some of the time also use a screwdriver for other purposes: perhaps as a poker, pry bar, nail set, cutting device, staple remover, or tool to dig up weeds. Third, technology changes through an iterative interactional process whereby it gets invented and then redesigned based on user interactions; it then is reintroduced to users, further interactions occur, and further redesigns implemented. Typical software updates (e.g., of word processors, browsers, and operating systems) epitomize this iterative process. Fourth, a society can reject any technology, or delay its acceptance. The Chinese, for example, invented important technologies, such as gunpowder, which then transformed medieval Europe, but which the Chinese themselves did not readily adopt, likely because they were seeking to slow the rate of change. Finally, fifth, no society or even subculture is completely homogeneous, and thus the adoption process can differ across groups. Orlikowski (2000), for example, studied the use of Lotus Development Corporation's Notes software by two groups within a large multinational consulting firm: technologists and consultants. The technologists used Notes extensively. They used e-mail, maintained electronic discussions with Notes databases, and created their own database designs. In contrast, the consultants used Notes minimally, sometimes even begrudgingly. The consultants had doubts about the value of Notes for their performance, and they were also under a time-based billing structure. "Because many consultants did not see using *Notes* as an activity that could be billed to clients, they were unwilling to spend time learning or using it" (p. 416). The consultants also feared that the collaborative properties of Notes would threaten their status within the company. Thus Orlikowski proposes an interactional "view of technology structures, not as embodied in given technological artifacts, but as enacted by the recurrent social practices of a community of users" (p. 421).

In chapter 2 I argued against genetic determinism. I said that although we are shaped and constrained in important ways by our evolutionary heritage, an essential aspect of our human nature is that we have choice, autonomy, and free will. Here I am making a similar argument against technological determinism and reinstating the primacy of the individual and the role of culture.

Designing Technology for People to Flourish

Several long-standing views of technology emerge from Fritz Lang's 1927 silent movie *Metropolis* (Pommer). In this movie, the wealthy planners, technologists, and thinkers live above ground, in light, while the economically impoverished workers live and toil below ground. The enormous factory machines demand harder and faster work. When workers die, to keep pace, more workers are brought in. The workers consider starting a revolution. But a beautiful woman, Maria, steps forward and argues forcefully that no matter how desperate the workers rightly feel, another path should be followed: to wait for "The Mediator." It is this entity, Maria says, who will unite the two halves of society, who will be the "heart" between the "head" (the higher class) and the "hands" (the lower class). As the plot unfolds, a mad scientist makes an evil robot, in the form of Maria. The biological good Maria is imprisoned. The robotic evil Maria (who becomes unleashed from its programming, and becomes conscious and autonomous) then foments the workers' unrest and instigates a revolution. The workers begin to destroy the technological underpinnings of the city. Then, as the city starts to flood, the workers realize that the technology sustains the lives of their children and wives, and themselves. The robotic evil Maria is discovered, and captured and burned at the stake. The biological good Maria escapes from prison and helps save the children and city from destruction, and again holds out the possibility of uniting the head and the hands.

We of the twenty-first century are not the first to recognize the downsides of technology. We are not the first to be fascinated with creating artificial life. And we are not the first to hold out a vision of technological humanism: technology by which people can flourish. How can this vision be achieved?

Over the last decade, toward addressing this question, I have assisted my colleague, Batya Friedman, in her pioneering work in what she calls *Value Sensitive Design*. I would like to draw from some of our published papers to provide a brief synthesis of this design method (Friedman and Kahn 2008; Friedman, Kahn, and Borning 2006; Friedman, Kahn, et al. 2006; Friedman, Kahn, and Howe 2000; Friedman, Smith, et al. 2006). By doing so, I set in motion my specific studies on technological nature, which comprise the empirical core of this book.

Value Sensitive Design

Value Sensitive Design is a methodology that helps people design technology to support human values in a principled and comprehensive manner throughout the design process. The central human values focused on, so far, include human health, privacy, freedom from bias, universal usability, trust, autonomy, informed consent, accountability, and environmental sustainability.

Value Sensitive Design maintains that human values exist and have moral standing not simply because of personal interests or cultural standards, but through the construction of knowledge, universally shared. That, of course, is a contentious claim, and it has engaged a good deal of debate (Friedman and Kahn 2008; Friedman, Kahn, and Borning 2006). At the same time, Value Sensitive Design maintains that how such values play out in a particular culture at a particular point in time can vary, sometimes considerably. For example, some technologists have argued that privacy is a quaint concept that we need to let go of, because soon there will not be any, given technology's ability to track the flow of each human life and link databases. In contrast, Value Sensitive Design posits that some degree of privacy is a universal good. It is necessary, for example, for a healthy child's and adolescent's construction of self and identity, and it remains important throughout the human life span. At the same time, Value Sensitive Design recognizes that how privacy plays out across cultures varies tremendously. For example, in many

Western homes, to establish a private space a person might retreat to an individual bedroom. In a traditional Inuit igloo, to establish a private space, a person might turn one's face to a wall, with the expectation that others in the igloo would then not look in that person's direction (Houston 1995). Accordingly, Value Sensitive Design maintains that privacy needs to be protected by designing the right kinds of technological systems for particular cultural contexts.

Methodologically, Value Sensitive Design integrates and iterates on three types of investigations: conceptual, empirical, and technical. In brief, *conceptual investigations* comprise analyses, often philosophically informed, of the central constructs and issues under investigation. For example, whose values should be supported in the design process? How are values supported or undermined by particular technological designs? Might there be underlying biological constraints in how humans can use or adapt to particular technologies? How should we engage in trade-offs among competing values in the design, implementation, and use of information systems (e.g., autonomy vs. security, or anonymity vs. trust)? *Empirical investigations* focus on the human response to the technology, and on the larger social context in which the technology is situated. Accordingly, the entire range of quantitative and qualitative methods used in social science research may be applicable here, including observations, interviews, surveys, focus groups, experimental manipulations, measurements of user behavior and human physiology, contextual inquiries, collection of relevant documents, and heuristic evaluations. Finally, *technical investigations* focus on the design and performance of the technology itself. Here it is assumed that technologies provide value suitabilities that follow from properties of the technology. For example, an online calendar system that displays individuals' scheduled events in detail readily supports accountability within an organization but makes privacy difficult.

To illustrate Value Sensitive Design in process, consider an early project by Friedman, Felten, and colleagues as they sought to understand how to design Web-based interactions to support informed consent, particularly through the development of new technical mechanisms for cookie management in a Web browser (Friedman, Howe, and Felten 2002; Friedman, Millett, and Felten 2000; Millett, Friedman, and Felten 2001). Friedman et al. began their project with a conceptual investigation of informed consent. They drew on diverse literature, such as the Belmont Report (U.S. Department of Health, Education, and Welfare [DHEW], 1978), which delineates ethical principles and guidelines for the protection of human subjects, to show that the concept of "informed" encompasses disclosure and comprehension, whereas the concept of "consent" encompasses voluntariness, competence, and agreement. Each of these subcategories in turn revealed rich philosophical distinctions. To validate and refine their resulting conceptual analysis and initiate their technical design work, Friedman et al. then conducted a retrospective analysis of how the cookie and Web browser technology embedded in Netscape Navigator and Internet Explorer changed—with respect to informed consent—over a five-year period, beginning in 1995. This investigation of each browser's technological properties and underlying mechanisms led them to conclude that, although cookie technology has improved over time regarding informed consent, some startling problems remain. For example, as of 1999, in both Netscape Navigator and Internet Explorer, the information disclosed about a cookie still did not adequately specify what the information would be used for or how the user might benefit or be harmed by its use. Moreover, the default setting for both browsers was to accept all cookies with no obvious visibility to the user. Friedman et al. then used the results from the above conceptual and technical investigations to guide their redesign of the Mozilla browser (the open-source code for Netscape Navigator). Specifically, they developed three new types of mechanisms: (a) peripheral awareness of cookies; (b) just-in-time information about individual cookies and cookies in general; and (c) just-in-time management of cookies. They periodically conducted formative evaluations of their work in progress to assess how well their design supported the user experience of informed consent. Their assessment instruments for informed consent also drew from and later provided guidance to their conceptual investigation. For example, during one of the initial empirical investigations, Friedman et al. discovered that users wanted to control cookies with only minimal distraction from their task at hand. This finding not only contributed to the technical designs described above (which incorporated peripheral awareness and just-in-time interventions), but also enhanced the initial conceptual investigation (by adding the criterion of minimizing distraction from the task at hand).

One last technique of Value Sensitive Design is worth emphasizing, as its societal implications are large, and I employ it in my studies on technological nature. It is to take seriously two classes of stakeholders: direct and indirect. Direct stakeholders refer to parties—individuals or organizations—who interact directly with the technology or its output. Indirect stakeholders refer to all other parties who are affected by the use of the technology. Often, indirect stakeholders are ignored in the design process. For example, computerized medical records systems have often been designed with many of the direct stakeholders in mind (e.g., insurance companies, hospitals, doctors, and nurses), but with too little regard for the values, such as the value of privacy, of a rather important group of indirect stakeholders: the patients.

Conclusion

This chapter contains four overarching ideas, which, in the order I presented them, I think of as speculative, obvious, nuanced, and novel. For the speculative, I started with an account of why the technological turn occurred. The account might be right. If it is, it points to the primary position of natural resources and a secondary position of citadels and then cities in explaining how we became a technological species. If it is wrong, then at a minimum few will disagree that the technological turn happened. Others speak of this event in evolutionary history as the "great leap forward" (Diamond 1997; Ehrlich and Ehrlich 2008). But to my mind that metaphor conveys a straight evolutionary path, like an Olympic long-jumper who runs straight and fast and then leaps forward, whereas the metaphor of turning helps to convey the idea of changing the course of human life, which technology did.

My argument for something gained, something lost, seems to me so obvious that I hesitated including it. Yet I sense I am living within a global culture that views most technology as unambiguously good. The worldview seems to be that if the technology can be designed and built, go for it; if there are problems, we can figure it out later, at which point, likely enough, a new technology will solve the problem. Mumford (1961) says that this view has been around for fifteen thousand years. Of course, it does not work that way. As Diamond (2005) points out: "New technologies, whether or not they succeed in solving the problem that they were designed to solve, regularly create unanticipated new problems" (p. 505). In addition—and this is what I want to highlight—through technological innovation we lose a previous form of experience, often a more natural form of experience, sometimes of a form that is better for us physically and psychologically. Thus, by focusing on something gained, something lost, we position ourselves to make better choices about which technologies to buy as consumers, which technologies to adopt as a culture, and which technologies to develop as designers and engineers.

My empirical studies on technological nature, in chapters 4 through 10, build on this general orientation toward understanding psychologically (and physiologically) what we gain and what we lose when technologies mediate, augment, or simulate nature. These studies, as the reader will recognize, are grounded by the nuanced interactional position on how technology affects people, and are structured by the novel design methodology, Value Sensitive Design, which I have described above. We are now in position to move to the empirical studies themselves.

4
A Room with a Technological Nature View

The story of Michel Faber's (2005) "The Eyes of the Soul" begins with Jeanette and her son looking through their apartment window onto the graffiti-filled concrete walls of a small grocery shop in their blighted urban neighborhood. Outside the shop, unsavory parts of the neighborhood congregate. They see kids sniff glue and drug users argue with the police. They see violent gangs acting violently. Jeanette once looked out the window and saw "a drunken, bloodied boy larking about on the roof of the shop, pissing over the edge, while his mates whooped and ran around below, dodging the stream" (pp. 45–46). Jeanette had tried pulling the blind down over the window. But that made her apartment feel "like a prison" (p. 42), and she figured it was better to see out onto anything than not to see out at all.

One day a saleswoman knocks on Jeanette's door. The saleswoman is gentle, direct, effective: "We offer people an alternative to windows" (p. 39), she says. As the saleswoman talks, her assistant begins mounting a large dull-gray screen onto the outside of the apartment window. Then the saleswoman hands Jeanette the remote, and tells her to switch it on. "'Switch it on?' echoed Jeanette. 'Yes,' said the saleswoman, nodding encouragingly as if to a small child. 'Do go ahead. Feel free'" (p. 42). Jeanette switches it on. The house then seemingly relocates itself into the middle of a spacious country garden. "There were trellises with tomatoes growing on them, and rusty watering cans, and little stone paths leading into rosebushes, and rickety sheds half-lost in the thicket. . . . At its wildest peripheries the garden merged (just about at the point where the Rusborough shop ought to be) into a vast sloping meadow that stretched endlessly into the distance. . . . In the sky above, an undulating V-formation of white geese was floating along, golden in the sunlight"

(p. 43). Jeanette moves closer to the window. She sees the smudges that have been on the glass for at least weeks. Through the window, "the world really was what it appeared to be, radiant and tranquil. The perspective changed subtly just the way it should, when she turned her head or looked down" (p. 43). "Her ear, so close now to the glass, heard the little beak of that sparrow quite clearly, the infinitely subtle rustle of the leaves, the distant honking of the geese" (p. 44). Shakily, Jeanette asks the saleswoman whether it's a video. It's not. Jeanette says it then must be some kind of film and asks how long it plays for. The saleswoman says, "It goes *forever* . . . . It's a real place, and this is what it's like there, right now, at this very moment" (p. 44). Though she has limited income, Jeanette signs the monthly contract for the new window. "She knew she was making the right decision, too, because while the screen was being bolted onto her house, it had to be switched off briefly, and Jeanette missed her garden with a craving so intense it was almost unendurable" (p. 47). Later that day, when her son came home: "He pointed, unable to speak. Finally, all he could manage was: 'Mum, what are those birds doing there?' Jeanette laughed, wiping her eyes. . . . 'I don't know,' she said. 'They're just . . . they just live here'" (p. 48).

Faber offers us an instantiation of technological nature: a technological nature window that transforms the ugly view of urban blight into a beautiful experience of nature outside one's domicile. The view is not a recorded video but actual nature displayed in real time. If the geese decide to land or fly away, that is what Jeanette will see. The sun rises and transects the skyline in accord with the season's trajectories and daily modulations of weather. A key point of the story is that this instantiation of technological nature "works" psychologically: Jeanette's emotional life is enhanced. She experiences nature's beauty. She feels happy, joyful, fulfilled.

Faber's story is one of fiction. But what if something like Faber's technological nature window actually existed—would it lead to improved human well-being?

My colleagues and I investigated this issue empirically (Kahn et al. 2008). We created a technological nature window by installing a large plasma screen in an office, and on it we displayed production-quality HDTV real-time local nature views. Through this installation, we sought answers to the following questions: In terms of physiological recovery from low-level stress, would people be more restored by this technological nature window view compared to having no window view at all? Would people be more creative when working in a room with a technological nature view compared to no view? Would they enjoy it more? In addition—and this is a critical question that Faber does not pose in his futuristic story—how would the physiological and psychological effects of viewing nature through this technological nature window compare to viewing the same nature scene through a transparent glass window?

Toward answering these questions, we conducted an experiment where we assigned 30 participants to each of three office conditions. Each condition employed the same office on our university campus. The office size was approximately 13 feet by 8.5 feet, with off-white colored walls, matte finished, and 10.5 foot ceiling with fluorescent ceiling lights. All 90 of the participants were undergraduate students (age 18–34; mean age, 20–28). The three conditions were as follows.

The Glass Window Condition

In the first condition, the office had a view through a glass window that overlooked a nature scene that included water in the foreground, as part of a public fountain area, and then extended to include stands of deciduous trees on one side, and a grassy expanse that allowed a visual "exit" on the other. This office view was chosen to include features that people usually find aesthetically pleasing and restorative in nature (Kaplan and Kaplan 1989; Orians and Heerwagen 1992). An office desk, 7.5 feet by 2 feet (32 inches high), was placed in front of the window. A swivel chair was locked into position on the floor so as to keep constant the distance to the window.

The Blank Wall Condition

In the second condition, we sealed off the glass window with light-blocking material, and then covered the sealed window with drapes, in effect turning the space into a windowless office.

The Technological Nature Window Condition

In the third condition, we inserted a 50-inch plasma screen into the office window, entirely covering it (see figure 4.1). We then mounted a

Figure 4.1
Demonstrator in the technological nature window condition. The technological nature window covered up the same-sized window used in the glass window condition. The camera that recorded looking behavior can be seen poking out from the drapes to the left of the technological nature window. The drapes were pulled across the entire wall for the blank wall condition. During the actual experiment in the glass window and technological nature window conditions, the fountain at the center of the body of water was splashing water into the air, and thus was more dynamic than is conveyed in this photograph.

production quality HDTV camera approximately 15 feet higher on top of the building and, through hard cabling, displayed on the plasma screen essentially the same glass-window view one would see from inside the office itself. The size of the glass window and the technological nature window were virtually identical. The desk was moved approximately four feet from the technological nature window so as to mimic as closely as possible from a viewer's perspective the experience of the same view through the glass window.

Notice that our technological nature window (given current technology) was not as advanced as Faber's futuristic window, especially insofar as ours did not solve the parallax problem, meaning that the view did not shift as one moved around the screen. It will likely be decades until the parallax problem is solved in a way that even approaches Jeanette's seamless experience in Faber's story. Nonetheless, our technological nature window, like Faber's imaginary one, displayed in real time a view of nature.

At the start of the experiment, which lasted approximately 1.5 hours for each participant, a researcher attached disposable electrodes to participants' lower ribs to capture heart rate. (For readers with more specialized physiological knowledge: we used a Biopac MP 100 physiological system with a two-lead configuration to collect electrocardiogram waveform data at a rate of 200 samples per second. Cardiac interbeat interval was determined from the electrocardiogram waveform based on the interval between R-waves, and heart rate was computed as the reciprocal of interbeat interval.) Participants then had a five-minute "waiting period" during which they could stare out the glass window, at the technological nature window, or at the blank drapes, if they so chose. Next, participants completed a series of four tasks: a proofreading task (to assess cognitive focus), a "tin can unusual uses" task that asked for different uses for a tin can (to assess divergent creative thinking), a "Droodle" task that asked for a clever label for an ambiguous drawing (to assess linguistic creativity in relation to a figurative form), and a "Modified Droodle" task that asked for the creation of one's own Droodle (to assess both linguistic and figurative creativity). I will say more about each of the three creativity tasks in the section on creativity results to follow. Tasks were counterbalanced, except that the Droodle always preceded the Modified Droodle (since the former helped participants understand the latter). Following the tasks, participants had another five-minute waiting period. During all this time, a camera (visible in figure 4.1) recorded participants' eye movement. The camera was time-synchronized with the physiological recording equipment. Following completion of the second waiting period, all electrodes were removed. We then engaged each participant in a 50-minute social-cognitive interview about their judgments and reasoning about windows.

I report now on what we found. The results are grouped by our three areas of investigation: participants' physiological restoration, creativity, and thoughts about windows. The first group of results on physiological restoration has been reported more formally and extensively in Kahn et al. 2008, and the second and third groups of results are being reported here for the first time.

Physiological Restoration

Prior to the start of each of the six activities (the two waiting periods and four tasks), a researcher gave each participant instructions for the new activity. This form of interaction typically elevated participants' heart rate, and thus functioned as a low-level stressor. To assess stress reduction, we computed the rate of heart rate recovery over the course of the next minute that followed each low-level stressor. The statistical analyses were sophisticated and extensive, and we refer the reader to our published research paper for details (Kahn et al. 2008). Nonetheless, our two key results are easy to summarize. First, there was more rapid heart rate recovery in the glass window condition than in the blank wall condition. Second, there was no difference in the heart rate recovery between the technological nature window condition and the blank wall condition.

Our first result—that being in an office that afforded a view of nature through a glass window was more effective in reducing stress than being in an inside office—is in line with Ulrich's (1984) classic study on the restorative benefits of viewing nature through a glass window. In that study, Ulrich examined the well-being of patients after gall-bladder surgeries who were assigned to one of two rooms. One room looked out onto a nature setting (a small stand of deciduous trees in a parklike area) and another room looked out onto a brown brick wall. Ulrich found that patients who had been assigned to the room with the nature view recovered better (e.g., they tended to have lower scores for minor post-surgical complications, required less potent painkillers, and had shorter postoperative hospital stays).

Our second result—that the technological nature window view was no more effective in reducing stress than a blank wall—surprised us. We had expected that this instantiation of technological nature would have garnered some, though not all, of the restorative benefits of the glass window.

It could be suggested that these results occurred because participants looked out the glass window more often than they looked out the technological nature window, and that physiological restoration depends on actual looking behavior. To address this possibility, we conducted a second-by-second coding of what each participant did with his or her eyes. Our results showed that participants looked just as often at the technological nature window (median, 58 looks per participant) as at the glass window (median, 52 looks per participant). However, the total duration of looking time was significantly greater in the glass window condition (median, 622.0 seconds) than in the technological nature window condition (median, 491.5 seconds). In other words, both windows just as frequently garnered participants' attention, and on this measure our results showed their equivalent functionality. But the glass window view held participants' attention longer than the technological nature window view.

Thus, based on this analysis, it could still be suggested that the physiological restoration in the glass window condition was due to longer looking time out the window. Accordingly, we next examined the rate of physiological restoration when participants were actually looking out each window. We found that when participants spent more time looking at the glass window, their heart rate tended to decrease more rapidly, which was not the case with the technological nature window. In other words, when we controlled for "eyeballs on stimulus," the glass window was more restorative than the technological nature window.

The Possible Impact of Other Variables

Our experimental design did not directly control for outside weather conditions, light intensity, and natural light. Thus, we conducted additional analyses to explore the possible impact of these variables on heart rate recovery.

Outside Weather It could be suggested that our physiological findings are an artifact of different weather that participants viewed in the glass window condition compared to the technological nature window condition. Thus, we collected independent weather logs from NOAA for the start and end of each experimental session. In brief, our resulting analyses showed no significant difference in weather conditions across the three experimental conditions. Furthermore, there was no evidence that the average of the heart rate slopes differed significantly by outside weather condition, either with all three experimental conditions combined or within each condition separately.

Light Intensity   It could also be suggested that our physiological findings are an artifact of what might have been greater light intensity in the glass window condition compared to the two other conditions. Indeed, we collected illumination data for each experimental session, and it was the case that in the glass window condition light intensity was much greater and more variable (median, 5,010.6 lux, range 277.7 to over 21,517 lux) than in either the blank wall condition (median, 456.4 lux, range 418.7 to 480.1 lux) or plasma window condition (median, 462.8 lux, range 338.0 to 477.9 lux). Accordingly, we examined the association between light intensity and heart rate recovery within the glass window condition. Results showed a nonsignificant correlation. That is, in the glass window condition, participants who experienced greater natural light did not experience greater physiological recovery. Thus, though not by itself conclusive, this result provides some evidence that light intensity by itself would not explain the differences in heat rate recovery between the glass window condition and the two other conditions.

Natural Light   Other research has suggested that people may accrue physiological and psychological benefits simply by experiencing daylight in otherwise inside spaces (Küller and Lindsten 1992; Leather et al. 1998; cf. Küller and Wetterberg 1993). Thus it could be argued that the effects of our current study could be completely explained on the basis of daylight: that the glass window condition was the only condition of the three that had actual daylight (as opposed to digitally represented daylight through the technological nature window). It is possible that the natural light played a role in our findings. But if it did, it was not the only variable. Recall that when we examined the relationship between time spent looking at the window and heart rate recovery, we found a significant association in the glass window condition but not in the technological nature window condition. In other words, actually looking out the glass window played a significant role in heart rate recovery. It is also possible that for our purposes the initial question is moot. Whether physiological recovery is enhanced by an actual view of nature or actual access to natural light, or both—all of these results would demonstrate that physiological recovery is enhanced by access to actual nature as opposed to technological nature.

In brief, outside weather was not a factor, and it does not appear that our results can be explained away by differences across experimental conditions in either light intensity or natural light.

Creativity

One impetus for investigating creativity in this study came from my own experience of writing. I feel more creative when I write with a view of nature through a glass window. It is not that I am always looking out the window; most of the time I am focused on paper or on my computer screen. But every few minutes my gaze drifts to the view. I mentally rest for a few seconds, and sometimes ponder. Perhaps every 5 to 15 minutes I look out the view for about a half-minute or more, and then I am back to writing. I have mentioned my experience to various colleagues. Some have similar experiences, and have pointed to the literature on attention restoration theory (Kaplan and Kaplan 1989), whereby it has been proposed that nature views have properties that engage involuntary yet undemanding attention, and thus promote recovery from mental fatigue, and potentially encourage the creative mind. Other colleagues have said that they personally find windows a distraction, and that they do their best work without one.

Thus, we sought to explore the possible role of creativity across the three experimental conditions. We were interested in two key questions. Are people more creative when working in an office that has a glass window view of nature compared to an inside office? And, in terms of promoting creativity, how does a glass window view of nature compare to a technological nature window?

The present experimental design allowed us to investigate these questions. Our methods, however, were not altogether straightforward insofar as we could find no single established measure for creativity. Perhaps that is not surprising insofar as one can presumably be creative in many different ways, for example, linguistically, musically, mathematically, and humorously. But even within a single domain, we did not find any ideal measure. Thus, our approach was to use one established measure (the Unusual Uses Task) and two measures that we created (the

Name a Droodle Task and the Invent a Droodle Task). I report on our tasks for two reasons. First, the tasks themselves have to date only been disseminated in a technical report (Kahn, Friedman, Severson, and Feldman 2005) and in themselves provide a contribution to the growing body of research on creativity. Second, given our results, it is important first to present enough detail to convey to the reader that our methods for assessing creativity were thoughtful and varied, and that the data were analyzed carefully.

The Unusual Uses Task
In this task, we gave participants 10 minutes to write down as many possible uses of a tin can as they could think of. We then scored the resulting data in two ways. One has been standard in the creativity field (Torrance 1962) and is based on the number of items generated, with a higher total representing higher creativity. However, that method seemed to us somewhat flawed, as one could generate many uses of a tin can that are all virtually the same (e.g., use a tin can as a "container for peas," "a container for corn," a container for carrots," and so on), which seems less creative than generating distinctly different uses (e.g., "throw rocks or marbles in it," "slug habitat," "chimney starter for coals," and so on). Thus, second, we created a new system for coding the Unusual Uses Task that accounted for seven different creative strategies, which we characterized as follows: (1) *Transformation*, wherein the tin can is changed in form or structure (e.g., "melt down to make statue"); (2) *Augmentation*, wherein the tin can is in some way enhanced (e.g., "roller for hair, heat and use as curling iron"); (3) *Categorical Reduction*, wherein one aspect of the tin can is used (e.g., "used to cut round for biscuits"); (4) *Experimental Medium*, wherein the tin can is used as the means to explore the environment (e.g., "rain/snow gauge, to see how much rainfall there was"); (5) *Proliferation*, wherein the tin can is changed into many tin cans (e.g., "dangle a bunch of cans from string to make a wind chime"); (6) *Pretense*, wherein one overlays subjective mental constructions that involve persona, agency, function, or attributes onto the tin can (e.g., "rubbing it to find genie inside"); and (7) *Symbolic Abstraction*, wherein the tin can is conceptualized as representing a higher-order concept, principle, idea, or class (e.g., "religious icon").

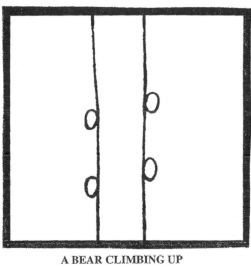

**A BEAR CLIMBING UP
THE OTHER SIDE OF A TREE**

Figure 4.2
An example of a Droodle.

The Name a Droodle Task

A Droodle is a simple abstract drawing that "comes into focus" (in a surprising way) with the addition of a clever title. Figure 4.2 illustrates an example from Price, Lovka, and Lovka (2000).

To our minds, creative Droodles depend primarily on people's ability to use language in witty and creative ways around an image or figure. Other tests of creativity have similarly assessed language. For example, Guilford developed a task in which participants had to create clever titles for story plots (Torrance 1962).

In our experiment, we explained to participants what a Droodle was, and provided them with some examples. Then we presented them with four new Droodle drawings, selected from the collection by Price, Lovka, and Lovka (2000). We told participants that they had three minutes to generate the best Droodle descriptions that had an element of surprise.

We then coded participants' Droodles into four categories: (1) Non-Droodle, which refers to Droodles (a) that lack a title, (b) where the title lacks coherence or relevance with regard to the drawing, or (c) where the title is only marginally relevant to the drawing and does not depend on the intentional creation of the title at hand (i.e., any randomly generated title could possibly have marginal relevance to the drawing). (2) Low Droodle, which refers to Droodles that (a) are literal in that the title provides a satisfactory description of the drawing, (b) attempt to convey a Droodle strategy without success, or (c) lack coherence. When viewing the Droodle, the response is more of a "blah." (3) Medium Droodle, which refers to Droodles that (a) are somewhat literal but successfully convey a Droodle strategy, or (b) are somewhat abstract and successfully convey a Droodle strategy, but lack the luster or coherence of a High Droodle, or may lack coherence. When viewing the Droodle, the response is more of an "aha, but . . ." or "okay, sure." (4) High Droodle, which refers to Droodles that (a) are abstract and successfully convey a Droodle strategy, (b) result in a shift or twist in how one would view the Droodle (there may be a literal interpretation following the shift), (c) have a certain sophistication or surprise, or (d) have coherence. When viewing the Droodle, the response is more of an "aha."

Figure 4.3 consists of pictures of the four Droodles used in the experiment and examples of coded responses from the data that reflect our four levels of creativity.

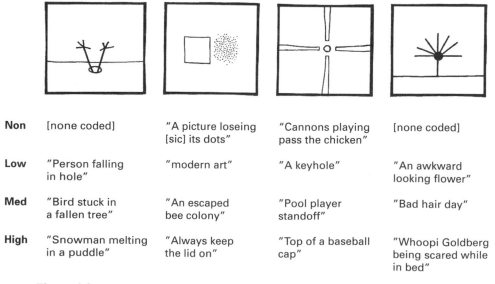

Non	[none coded]	"A picture loseing [sic] its dots"	"Cannons playing pass the chicken"	[none coded]
Low	"Person falling in hole"	"modern art"	"A keyhole"	"An awkward looking flower"
Med	"Bird stuck in a fallen tree"	"An escaped bee colony"	"Pool player standoff"	"Bad hair day"
High	"Snowman melting in a puddle"	"Always keep the lid on"	"Top of a baseball cap"	"Whoopi Goldberg being scared while in bed"

Figure 4.3
Four published figures of Droodles and coded prototypic written titles based on levels of creativity.

The Invent a Droodle Task

In this task, we modified the Droodle task, and asked participants to invent the best Droodle they could, both with a creative figure and title. We told them that they had seven minutes for this task. We coded their Droodles based on the same criteria established for the Droodle task. Figure 4.4 contains representative Droodles from the data representing the four levels of creativity.

With our methods established, we can now turn to our results. Recall that we had asked ourselves: Would participants be more creative in the glass window condition than in the blank wall condition? How would the technological nature window fare? Our results were not at all straightforward.

On the Unusual Uses Task, there were no statistically significant differences among the three conditions. Indeed, the median number of items generated per participant was exactly the same (median, 24 items) in all three conditions. Even when the results were broken down by the seven different creative strategies, there were still no differences across the three conditions in either (1) the percentage of participants who used each strategy at least once, or (2) the median number of times that participants used each strategy. Thus the conclusion could be that this form of creativity is not at all enhanced or diminished by our three different conditions: glass window, technological nature window, and blank wall. The caveat, however, is that in our directions to participants we had asked the following: "For this task, think of and write down as many different uses you can think of for a tin can. Write each use on the following sheet. You will have 10 minutes." As I noted earlier, however, these directions focus on quantity of production, not quality. We provided these directions because it was the standard way to administer this task, with the assumption by other researchers that more items meant more creativity. Once we recognized what we thought of as this error, we sought to correct it by generating a new system for coding the data based on the type of creative strategy used. But our new system—which we think of as in itself a contribution to the field—could not solve our initial mistake in how we asked participants to generate their items. In hindsight, we wished we had asked something like the following: "For this task, think of and write down as many highly creative uses you can think of for a tin can. Write each highly creative use on the following sheet. You will have 10 minutes."

Prototypic Non-Droodle

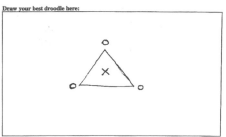

Description of Droodle: ~~Galaxy~~ ① The ~~animals~~ to fight for
getting ready   the food
② The new galaxy found to have 3 planets circling   in the center
Scratch Area – Sketch your ideas here   in a triangular path around the
center of the universe.

Title: "The animals getting ready to fight
for the food in the center. . . ."

Prototypic Medium Droodle

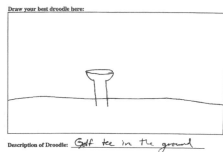

Description of Droodle: Golf tee in the ground

Title: "Golf tee in the ground"

Prototypic Medium Droodle

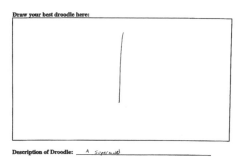

Description of Droodle: A supermodel

Title: "Supermodel"

Prototypic High Droodle

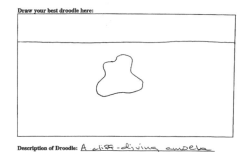

Description of Droodle: A cliff-diving amoeba

Title: "A cliff-diving amoeba"

Figure 4.4
Coded prototypic examples of Droodles (figures and written titles) based on levels of
creativity.

On the Invent a Droodle Task, participants had significantly higher creativity scores overall in the glass window condition than in the blank wall condition. Creativity scores for the technological nature window condition fell somewhere in between, but statistically were not significantly different from either of the other two conditions. In addition, though the result was not quite statistically significant, focusing only on participants in the glass window condition, those who spent more time looking out the glass window tended to have slightly *higher* scores on this task than those who spent less time looking out the glass window. Conversely, in the technological nature window condition, participants who spent more time looking at the technological nature window tended to have slightly *lower* scores on this task than participants who spent less time looking at the technological nature window (though again the difference was not quite statistically significant). These results support the proposition that this form of creativity is enhanced by an actual nature view compared to no view, and suggest (descriptively but not quite statistically) that the technological nature view comes up short in comparison to the actual nature view.

This account, however, gets particularly muddled when we look at the results from the Name a Droodle Task. In this task, participants in the blank wall condition overall had significantly higher creativity scores than participants in the glass window condition. The technological nature window condition again fell somewhere in between but was not quite significantly different from either of the other two conditions. One possible explanation for the higher creativity scores in the blank wall condition is that creativity in the Name a Droodle Task requires visual attention to the visual information of the Droodle itself, and the glass window acted as a "distracter." But this explanation was not supported when we analyzed looking behavior and creativity: We found—as we did with the Invent a Droodle Task—that participants who spent more time looking at the glass window tended to have slightly higher scores in the Name a Droodle Task than those who spent less time looking at the glass window.

In summary, there are some indications that actual nature views may enhance certain types of creativity. But for other types of creativity, the view may not matter. And for still other types of creativity, any nature view (actual or technological) may hinder creativity, though for those tasks, if one does have an actual nature view, then it still seems better to use it (look at it) than not. That is a confusing set of results. To know what is going on here—even to speculate intelligently—would seem to require further research. What might that look like? My inclination is that it would first be necessary to establish a wide range of valid creativity measures. That in itself could take a good deal of time. We started on that endeavor in this study. We did not get far enough. With a wide range of valid measures in hand, one would then be in the strong position of applying them systematically to this domain of inquiry.

Thoughts about Windows

As noted earlier, after the completion of the physiological part of this study, we engaged each participant in a semistructured interview about their judgments and reasoning about window environments. The interviews lasted about fifty minutes. We transcribed each interview, and then developed a system for coding the qualitative data. In what follows, I highlight some of the general trends. It is worth noting that our findings were largely the same whether or not participants experienced the technological nature window itself (i.e., had been assigned to that condition in the experiment).

The majority of participants (across the three conditions) had positive judgments about glass windows. Specifically, in their description of their ideal work space, the majority of participants mentioned unsolicited having a (glass) window (70%). The majority of participants also believed that every office should have a glass window (69%), and virtually all the participants said that if given the choice they themselves would choose an office with a glass window over an inside office (96%). Participants believed that working in an office with a glass window makes people physically healthier (71%) and more creative (93%), compared to an inside office. Participants also believed that having an office with a window conveyed social status (92%) and that such status was not arbitrary (in the social-conventional sense, for example, that "designer clothing" conveys status) but intrinsic in some way to the window itself (76%).

What were participants' reasons for their positive judgments about windows? Their reasons were varied and included the following considerations. Some participants spoke of glass windows affording the experience of outside light, air, and sound, and that it connected them to both nature and other people. For example, one participant said that with windows "I can see the weather and the people and feel the sun coming through; and I like that." Another said: "I can see the world as it really is; my mind responds to that . . . [for] what I'm doing actually is connected to the world." Another said: "I appreciate windows and the connection with nature, and you don't feel locked up all day." Some participants emphasized how a window provided them a means not just to receive the outside view, but to have an impact on what is going on outside. For example, one participant said: "I can just stick my head out the window, if I see somebody I recognize I can holler at them." Some participants spoke about how windows helped them in their work. For example, one participant said: "I've found that a little distraction is good for me, because I tend to be very, very intensely focused . . . and so if I can look out a window, it'll take me out of myself for a second and then I'll be refreshed to go back to it . . . a window kind of lets me balance those two things." Another said: "It's when you're like writing a paper and every once in awhile if you have like writer's block, just looking out the window kind of like helps you think and stimulates you." Some participants in the window-view condition spoke of feeling less stressed in the study itself because of the window. For example, one participant said: "[Looking out the window] I could feel all the tension dropping, my shoulders went down, I sat back, and my head kinda came up and just like all these muscles went [makes relaxing noise], and that's gotta be healthier in general, maybe it aids the flow of oxygen to the brain or something I'm not exactly sure what it was but I sure as hell felt a lot better."

In comparison to a glass window, participants were often less enthusiastic about a technological nature window. In one question, for example, we asked: "Let's say you lived in a culture where technology has a lot of status, like the United States. Would you rather have an office with a real window or with a high-tech video screen [technological window]?" Only 13 percent of the participants chose the technological window. In addition, only a minority of the participants thought that working in an office with a technological nature window would help people's health (39%), and the majority of participants said that there were downsides to having a technological nature window in windowless offices (81%). That said, if the comparison was to a windowless office, then participants were more enthusiastic about a technological nature window. Specifically, the majority of participants thought it would be a good idea to put technological nature windows in inside offices (72%), and that the technological nature windows would help people be more creative (76%) and better at problem solving (64%).

In their reasoning against a technological nature window, participants also articulated a wide variety of issues and concerns. Some participants spoke of the importance of sunlight and air, particularly in terms of human health. For example, one participant said that the technological nature window "is just a TV and you know it's not giving me natural light or fresh air or any of those things that makes us healthy." Another said that technological nature windows "make me jittery and they make me nervous, it's not the same as natural light and it's not the same as, oh God that would drive me nuts, oh God I just get a headache thinking about it." Some participants said that just because something is technological does not mean it is good for human health. For example, one participant said: "I just think about like in the fifties how TV dinners were cool and now it's like you should make your own food from scratch because it's healthier so just because it's from technology doesn't mean it's better for you." Some participants spoke against the economic costs of a technological nature window. For example, one participant said "the costs, the electricity . . . because I heard these screens are pretty expensive." Some participants said that a technological nature window lacked authenticity. For example, one participant said "just because the fact that you know it's trying to fake this kind of artificial feeling." Another compared the technological nature window to "special effects in movies; there's a lot of people that just feel movies are artificial even though you really can't [tell the] difference, but you know it's not real, like what you're looking at is not physically there." Finally, some participants said that they would feel manipulated by technological nature windows. For example, one participant said: "I would just really resent the idea that someone was trying to manipulate me into thinking that I was outdoors."

In brief, participants liked glass windows. Recall that virtually every participant would choose an office with a window over an inside office. In this respect, they were not unlike Jeanette in Faber's story, who preferred a window looking out onto virtually anything in comparison to not having a window at all. In addition, participants were less enthusiastic about a technological nature window in comparison to a glass window.

Conclusion

Michael Faber sidestepped an essential question. As recounted in the beginning of this chapter, in Faber's futuristic story Jeanette compared a beautiful technological nature view to an ugly glass view of urban blight and violence. The technological nature window "won": Jeanette paid her money and had it installed over her glass window. But the question that Faber did not ask was how the technological nature view would compare to an actual view of the same scene through a glass window. What is at stake is the question: Does it matter that nature is "real"? Or can we digitize nature and display it with psychological impunity?

Four clear findings emerged from our experiment. First, in terms of heart rate recovery from low-level stress, working in the office environment with a glass window that looked out on the nature scene was more restorative than working in the same office without the outside view (the blank wall condition). Second, in terms of this same physiological measure, the technological nature window was no different from the blank wall. Third, when participants looked longer out the glass window, they had greater physiological recovery; but that was not the case with the technological nature window, where increased looking time yielded no greater physiological recovery. Fourth, in terms of their thoughts about windows, participants liked the glass window, and were less enthusiastic about a technological nature window. In terms of creativity, the results were inconclusive.

Overall, then, here is my interpretation of the results. Even though a technological nature window might look like a window, have a view like a window, and be used by people as a window, it does not confer all of the physiological and psychological benefits of a glass window view of nature.

That said, I am aware that the technologist—and as I suggested in chapter 3, all of us are technologists—can offer a compelling rebuttal. It takes the following form: "Wait one moment! The problem is simply that our technology isn't yet far enough along. What we need to do, and what Faber envisioned in his story, is (a) to solve the parallax problem, meaning that the view should shift as one moves around the screen, (b) to allow one to hear nature sounds in accord with the view through the technological nature window, and (c) to make the technological window indistinguishable from a glass window. It's possible! Give my lab twenty million dollars and five years, and I'll create a technological window of the future that's just as good as the real thing." This form of a rebuttal is impossible to refute empirically insofar as it makes an empirical claim based on technology that does not yet exist. Nonetheless, as various forms of technological nature continue to be engineered, it is fair game to "put them to the test." But the question is, what test? What are the benchmarks against which we will hold the technologist's creation? As an analogy, imagine a runner who wants to be a great runner. How does she know if she is succeeding? For speed, she would typically ask: How fast can I run (over an established distance)? For distance, she would typically ask: How far can I run (and at what speed)? Then she would take those numbers and compare them to those of great runners. It is not enough to say that she has a beautiful runner's stride or that her parents think she is a great runner. Similarly, it is not enough to say that a technologist's version of technological nature looks great, or has such-and-such technical "specs," or that the technologist's buddies think it is a great technology. Rather, it has to "measure up"—and here is the key point—based not just on technical benchmarks but on psychological benchmarks.

What might such psychological benchmarks look like? The four investigations presented in this chapter offer a partial answer, because that is another way of understanding them. One benchmark is whether people are physiologically restored. That was assessed by heart rate recovery from low-level stress. A second is how people use the instantiation. That was assessed by frequency and duration of participants' eye gaze behavior. A third is whether people are creative while using the instantiation. That was assessed by the Tin Can Task, the Name a Droodle Task, and the Invent a Droodle Task. Finally, the fourth was how people consciously reason about the instantiation. That was assessed through a semistructured interview.

Getting the right set of psychological benchmarks—conceptualized robustly and accepted in society—remains a critical endeavor for engineering technological nature for human good.

5

Office Window of the Future?

I took Michel Faber to task in the last chapter. My reason was that in his story "The Eyes of the Soul," he had not asked his reader to consider how his futuristic technological nature window compared directly to a glass window view of the same nature scene. A plausible response could be something like: "Who cares? Get real. The point is that many people now and especially in the future will not have access to actual nature, so if in our technological nature window we can even somewhat successfully mimic the experience of actual nature, we'll be coming out ahead."

The point is well taken. It structured this second study.

My colleagues and I installed our technological nature window in seven inside offices of faculty and staff in the Information School at the University of Washington. We then employed a field-study methodology to explore the user experience of these technological nature windows (Friedman, Freier, Kahn, Lin, and Sodeman 2008).

The technological nature window was the same one described in the previous chapter insofar as it displayed, as the default, the same real-time view of the large fountain, deciduous trees, and an expanse of grassy area that was outside their building (see figure 5.1). There was, however, one important difference. We configured the window to allow participants to "switch the window off" or to switch it in other directions whenever he or she wanted by using it as a large computer display. They could, for example, use the window to display webcams worldwide, still photographs, personal files, or a PowerPoint presentation to students in their office. In this way, we sought to reconceive of the office window of the future, such that it could take on multiple functions at the user's discretion.

The questions that intrigued us and had prompted us to take on this study were these. Does the technological nature window increase people's satisfaction with their inside office environment? Is there a "novelty effect" such that over time people tire of looking out the window? Or do benefits actually increase as comfort with the technology increases? How do people integrate the technological nature window into their displays of work-related information? Does the window increase people's awareness of other people and of the larger university community? Does the window increase people's awareness of a sense of place and a sense of time? Do the natural aspects of the technological window (e.g., water, trees, and sky) contribute to a sense of physical and psychological well-being? Finally, are people concerned about the installation's possible impact on the privacy of indirect stakeholders—those individuals who in the course of their regular business on the university campus pass through the scene and have their images displayed in an inside office?

Our participants were seven university employees with inside offices: Nina (faculty, age 40–50); Barbara (faculty, age 45–55); Daniel (student services staff, age 25–35); Lara (student services staff, age 20–30); Brin (student services staff, age 25–35); Doug (faculty, age 35–45); and Sue (student services staff, age 20–30). Pseudonyms and age ranges are being used to enhance participants' privacy. Each participant determined where in his or her office the technological nature window would be installed.

We utilized a naturalistic field study methodology (McGrath 1995), and collected data for each participant over a 16-week period: 6 weeks with the inside office "as is," 6 weeks with the technological nature window, and 4 weeks following the removal of the technological nature window with the office returned to its original state. Across the 16 week period, each participant completed seven 30- to 45-minute semistructured interviews that focused on the participant's (a) impressions and use of the technological nature window, (b) perceived effects of the real-time views of the outdoor scene, (c) awareness of people walking through the plaza area (the indirect stakeholders), (d) assessment of the importance of the technological nature window, (e) intuitions about work performance and health, (f) social interaction related to the technological nature window, and (g) experiences, reflections, and comments on any other related topics. In total, 30+ hours of interviews were conducted

and yielded 652 pages of interview transcripts. In addition, each participant completed 10 work satisfaction surveys, 10 mood surveys, 10 office
perception surveys, journal entries, and responses to e-mail queries.
Because we had only two displays, data collection for the seven participants was staggered. The seven field studies were completed during the
span of one year.

We triangulated data sources (Krathwohl 1998; Pelto and Pelto 1978)
and conducted three types of overarching analyses. *Narratives*: We drew
on each participant's journal, e-mail responses, and seven interviews to
construct an integrative narrative describing each participant's individual
experience. *Themes*: Toward a more systematic and synthetic account of
the user experience across participants, we drew on the seven interviews
to develop a set of overarching themes. Then each interview was coded
for the presence or absence of each theme. Moreover, each time a theme
was identified, it was then coded with a valence (positive, neutral, or
negative) to reflect the user's judgment about its content. *Evaluations*:
Based on each participant's seven interviews, we conducted a quantitative analysis for a subset of the standard questions contained in the
interview. Questions included in this analysis conformed to a binary
response format (e.g., "yes/no"; "recommend/do not recommend").
A total of 41 questions were analyzed across the seven interviews.
Intercoder reliability was assessed statistically for the themes and
evaluations, and ranged from what is considered good to excellent.

Here is what we found.

Narratives

The narratives provide the reader with a sense of the complexity, depth,
and diversity of user experience. Because of space limitations, I present
only three narratives, those of Nina, Barbara, and Daniel.

Nina

Nina is a 40- to 50-year-old part-time lecturer. Her office is the only one
without a window along a long hallway. Therefore, she notices the
change from bright to dark as she enters her office. Because of her office's
small size and layout, Nina decided to place the technological nature
window on the wall behind her desk (figure 5.1).

Figure 5.1
Nina with her plasma display window highlighting the local plaza and fountain view.

Before the window was installed, Nina described her office as a "deprivation chamber" where "there is nothing to make [her] feel connected at all." Soon following its installation, Nina said that although the window didn't "agitate her," it didn't soothe her either. She also still expected it to display static images: "I find I look at the screen as a static picture, and then I'm somewhat startled to see people, objects moving. So initially, I am not using a window metaphor in my brain."

By the second week of using the technological nature window, Nina experienced a change. She wrote, for example, in her journal: "The screen feels much more window-like. I'm more aware it's dynamic." This idea was expressed even more fully in an interview during that week. She said: "I have gotten over the initial like 'it [the window] is a type [of] picture.' No. It's a live feed to me now. . . . I turn to look at [it], see what's going on outside and see people wearing shorts so I really do get a sense of 'this is what is going on outside.'. . . You know, now I'm looking at, I'm really looking out the window."

Once Nina adopted a window-like perspective of the large display, psychological benefits appeared to accrue. She said, for example: "With effort I can now find some peace in the room that wasn't there before. It's not pervasive, it doesn't completely change the nature of the office, it does help when I choose to. . . . I've been working at my desk with my back to [the window] and periodically I will just kind of stop and look at it, if only just for a moment to get a sense of 'oh yeah, outside. . . .'"

After six weeks of working with the technological nature window in her office, Nina said that it "has helped to temper some of that frustration and agitation" that she had felt in her windowless office. She also said: "I've felt like if I've wanted to at any moment I could sort of refer to, take a look at the screen and get a sense of it's nice out, it's not nice out, there are people out there, there aren't people out there. . . . It's also a reminder, it's like there is a world out there so . . . you want to get done with what you're doing because you want to get back out in the world."

Nina particularly came to value the "soothing" effects of the display and how it helped to focus her mind: "to help my thought process, to help my state of mind."

While Nina increasingly interacted with the display as a window, she also recognized its limitations, particularly on the dimension discussed in the previous chapter, that it did not allow for parallax (the apparent shifting of objects when viewed at different angles): "I probably would look out a [glass] window a little bit more [than the technological nature window] because you do have that opportunity to kind of move around and vary the angle of what you're looking at."

It took some weeks, but Nina also began to integrate the window into her workplace practice, using it as a computer display, and that took her by surprise: "The use of it [the window] in the course of a meeting was really surprising to me because that was not something I had anticipated at all. . . . I originally intended to just be displaying an image during a meeting. But in the course of the discussion and wanting to look something up that we could look at together, it led to through the rest of the meeting, sharing information up on the screen, instead of sharing it on my monitor here on my desk."

Connection to the outside afforded by the window also became a catalyst for interpersonal interaction "that wasn't there before." Such interactions particularly arose when special events occurred by the fountain. For example, Nina wrote the following in her journal: "There are some, I mean there were, at 5 P.M., 20 people swimming in the fountain, plus those being thrown in, and within minutes of turning the screen on,

I had 5 people in the office watching the antics, pulling other people in to see. Even [another faculty member with a window] came into this office to help provide commentary."

Such connections to local activities by the fountain helped provide Nina with a sense of motivation, of connection to the larger university community, and of why her work as a lecturer at the university is meaningful. For example, she wrote: "Graduation Day—what fun to have a 'view' of all the hustle: hustle out by the fountain while I get ready for the commencement ceremony. Everything is all neat and clean and the presence of all the families reminds us that something does come of all this!!"

Four weeks after the technological nature window was removed from her office, Nina reflected back on her experience. She said: "I miss having the ability to take time to just look at what's going on outside . . . to just watch the world outside and kind of shift your thinking a little bit. That to me was probably the greatest asset."

Barbara

Barbara is a 45- to 55-year-old full-time faculty member. She said that it is important to her to be engaged in outdoor activities, with her local community, and with the larger world, natural and social.

Part of what Barbara appreciates about glass windows is that they allow her to see that all is well, and from that sense of security she is able to move forward creatively with her work. In an early interview, for example, she said: "I really do think that's possible [that windows support better thinking]. It's the old connecting out there with the environment. It's sort of all is well so I can think. . . . There's a sense of solace and peacefulness. Things are OK out here."

Barbara describes her windowless office as a "pretty isolating office," "a real disconnect." And she makes a concerted effort "to have a sense of peace" by "bring[ing] things in that do connect it [her office] to nature, like the screensaver [of a nature scene]" and a full-spectrum lamp.

Barbara chose to install the technological nature window right above her desk. Within several weeks, she began to experience its window-like aspects. For example, she said that it allowed her to "reconnect," "to not feel so closed in," and to have a "sense of relationship to what's going outside of this office." Such connection helped provide the sense of security that she spoke of earlier. For example, she said that although she doesn't look out the window too often, "there's a comfort-zone in knowing it's there, and I think that's the difference. It's a peripheral knowledge of knowing it's there if I want it . . . and I think that is very supportive."

The technological nature window also provided Barbara with a sense of weather and time, which then influenced her work practice. For example, sometimes she looked out the window and saw that it was raining, and decided to work more before leaving the building. Other times: "Lunch! It's a beautiful day out so I think I want to go for a walk." One journal entry noted: "Beginning to darken with clouds at 3:30 PM. Item errands to run—Screen allows me to see that I need to get going." Barbara also considered "a sense of sunrise, sunset" basic to daily functioning, which she partly regained through the technological nature window.

Soon after its installation, and in the weeks that followed, Barbara began to display various webcam images on the display: notably of local traffic conditions, Mount Saint Helens, the Grand Canyon (see figure 5.2), and London. She particularly enjoyed the visual real-time contact with parts of the world that she had traveled to. In these ways, it is worth recognizing that Barbara created more interactions with the technological nature window than she could have with a glass window, garnering connections to both the local and the global.

Barbara was keenly aware that the technological nature images did not fully substitute for the experience of being elsewhere; but they were not too far off, either, especially given the importance Barbara placed on knowing that all is well elsewhere: "This window will take me anywhere in the world, but it won't let me smell anywhere in the world. You know what I'm saying? So, the image is still an image. It's not actually being there, but it is very close. I mean it is very good and it's a very safe way to look at what's going on in the world."

Finally, given the importance that Barbara placed on global awareness and visual access to remote locations, it is noteworthy that she judged the local scene as a little more important than a remote scene: "I think these two—the Grand Canyon and [the fountain plaza] are my favorites. Probably [the fountain plaza] more than anything just because it's where

Figure 5.2
Barbara in her office with the plasma display window (showing a webcam of the Grand Canyon).

I am right now . . . I think it connects me closer to what's going on immediately outside."

Daniel

Daniel is a 25- to 35-year-old student services staff person. His desk is situated in a central open office space that he shares with three other staff members, and which provides access to four other enclosed offices. Daniel is the first person people encounter when entering Student Services. Thus his job entails frequent social interaction, and diversified, if not stressful, responsibilities. He chose to place the technological nature window within view of his forward gaze, replacing a white board that had been on the wall (see figure 5.3).

Initially, Daniel thought that the main benefit of the technological nature window was simply one of aesthetics: "I use it as just something very, something pretty to look at. That's the main interaction with it." But then, as with Nina, Daniel came to recognize that the technological

Figure 5.3
Daniel working in the student services reception area and office with the plasma display window.

nature window allowed him to take short mental breaks that helped refocus his mind and to see his work in the larger university context: "It helps me to see the bigger picture of what I'm doing and how to respond to an e-mail; to not get so hooked and hung up on all the annoying little details of this particular person you're corresponding with. . . . Maybe it's like a metaphor of analogies: that I can see that person or that e-mail or that particular work in a greater context because I can look out and I can see other contexts."

In addition, Daniel thought the technological nature window eased the confinement of the office space: "It is a cramped space or at least it feels like [it is]. Last year, during the crunch of admissions, we would have typically 13 people in the office at one time and, for that space, anything that can make it expanded feel would help. Although the proportion of that screen, the screen looks much better in Brin's office than it does in ours. If it [the window] were even larger, if it took up the wall in a better way, I think it would make the space feel better."

Daniel was also critical of this new technological form. One of his critiques centered on the potential objectification and intrusion into the privacy of people in the plaza: "I've known about this project for, probably since maybe around when it started. And, you know, I knew that the camera was out there. I never really thought about it. But when we installed the screen in our office and I saw that we could actually see the data that was coming through, it really, it started making me be very, feel very self-conscious around that fountain. And so it does kind of bother me."

The interviewer then asked Daniel if he thought it bothered other people. Daniel responded: "If they were aware of it? Yeah, I think it would. I think it would bother them, the fact that they aren't made aware of it. . . . It's some connection between the idea of technology and how it intrudes upon the ability to live your life without being scrutinized, without being moderated, without all these socially constructed technology evil things! [Daniel laughs.] Not quite that bad, but . . . it is an unease with technology and its intrusion into my personal and private time. . . . If someone is just looking out the window, all they have to rely upon is their memory, all we have to rely upon is subjective human experience, but technology somehow objectifies me in my life."

Moreover, Daniel worried that people are "short-sighted" when adopting technology, unaware of its effects and passing it to newer generations as a given. "If we have this kind of thing in every single office on every city block in America, yes, the people, the younger generations will be used to it; but that doesn't mean it's okay. Just because you're used to it doesn't mean it has to be that way. We don't have to be videotaped twenty-four hours a day. We don't have to always be . . . accessible to the public, you know? Just think about for example the telephone. It's changed how we live and how we interact with the world. Just think, prior to the telephone, there was public space and there was private space . . . there were public places to meet where you didn't have to pay to go to and now if you want to meet with someone . . . typically you'd have to meet them at a café. . . . And it's very hard for people to imagine that there used to be a time where you could meet with people at the village square. . . . Now we move into a world where we're videotaped at all times, what kind of things will we, they take as givens that don't need to be taken as givens. . . . What society considers acceptable privacy now will be considered too stand-offish."

Daniel's ambivalence—on the one hand, his attraction to the technology and its aesthetics, and on the other hand, his critique of the way in which technology unnecessarily objectifies people and intrudes into personal and public spaces—continued throughout the entire time he spent with the technological nature window, and into his final interview a month after the window was removed: "There's part of me that's missing the niceness that came with it, but another part of me is quite alright with having that lack because . . . I do not want it to be the norm, to be expected that we should get used to the idea that people in offices across the world may be watching me out there."

Themes

The above narratives provide knowledge of each participant's unique way of interacting with, conceptualizing, and judging the technological nature window, within the context of each participant's daily work environment. Moreover, the narratives highlight how each participant's interactions, conceptualizations, and judgments solidified or changed over a period of six weeks, which is a reasonably long period of time, especially as compared to the length of most psychological experiments that come in at under two hours, usually.

Still, from the narratives it is difficult to grasp the key findings within and across participants. This limitation occurs in part because of the obvious reason that I have reported on only three of the seven narratives (for considerations of space). But it also occurs for two other reasons. One, the narratives are lengthy and by nature difficult to distill. And, two, the narratives are somewhat idiosyncratic insofar as they do not formally seek to account for the presence or absence of central themes across participants. To address this limitation, we first identified ten central themes from the data. We then used these ten themes to characterize the pattern of user experiences for all seven participants. The ten themes are as follows.

Experience of Technology

This theme refers to the user's characterization of the technological nature window including references to the quality of the image, dynamic nature of the screen, cost of the technology, preoccupation with gadgetry, malfunction of the technology, and novelty (e.g., "The screen was flickering slightly this morning, and I turned it off during a meeting" [Brin, e-mail]).

Activities around Technology

This theme refers to the user's behavior in response to the technological nature window including references to work practice, decorative actions, physical movements, and rituals or routines (e.g., "realizing how ingrained it was becoming in my natural patterns, the number of times I'm looking to where it was and having that, 'Oh, it's not there anymore'" [Nina, interview]).

Sense of Place

This theme refers to the effect of the technological nature window on the user's perceptions of the physical surroundings, including references to the office and outdoor environment (e.g., "There was a wonderful moment earlier today where it was sunny and rainy at the same time—a joy to see if not experience" [Doug, e-mail]).

Social Connection

This theme refers to the effect of the technological nature window on the user's relationships to other people, either reciprocally in the user's office environment or nonreciprocally through the live video feed (e.g., "I felt more connected to . . . the department" [Barbara, interview]).

Personal Interest

This theme refers to the user's emotional attachment to or general enjoyment of the technological nature window (e.g., "Um, I've really enjoyed [the window] . . . it makes things better" [Sue, interview]).

Cognitive Function

This theme refers to effects of the technological nature window on the user's cognitive processes, including references to focus, efficiency, and creativity (e.g., "I think . . . [the window] leads into thinking better" [Lara, interview]).

Psychological Well-Being

This theme refers to the effect of the technological nature window on the user's mental restoration, relaxation, and general morale (e.g., "I think it helped me think a little more positively . . . it had sort of more of a mood effect" [Lara, interview]).

Status

This theme refers to the user's perception of the technological nature window as signifying power or prestige (e.g., "It [the window] certainly could start to convey status, I mean I worked at _____ and the bigger the monitor you had the bigger, you know, the more important you were. So there's status I guess it can convey" [Nina, interview]).

Social Expectation

This theme refers to the user's recognition of conventional practices around the use of camera and display technology (e.g., "It wouldn't bother me if someone . . . could see me walking . . . by the fountain, you know what I mean, because that's sort of what happens with the webcam . . . it could capture me for a second" [Brin, interview]).

Ethical Issues

This theme refers to the user's recognition of ethical implications of the technological nature window for either direct or indirect stakeholders, including references to privacy, informed consent, security or safety, intellectual property, and objectification of individuals and/or their images (e.g., "It feels slightly voyeuristic watching folks going about their business through the monitor . . . it feels sort of Big Brother–like" [Daniel, journal]).

Table 5.1 characterizes the patterns of individual experiences of the technological nature window (WIN) by means of a timeline of themes for each participant. In reading table 5.1, the first column of the timeline (Install WIN) refers to themes from the interview conducted in week 7 when the technological nature window was first installed in the participant's office. The second column (During WIN) refers to themes from

Table 5.1
Timelines of Themes by Participant

	Theme	Install WIN	During WIN	Remove WIN	Post WIN
Nina	1. Exp. of tech.	−	+	n	n
	2. Act. around tech.	+	+	+	+
	3. Sense of place	+	±	±	±
	4. Social connection		+	+	
	5. Personal interest				+
	6. Cognitive function	±	±	+	±
	7. Psych. well-being	+	+	+	
	8. Status		+		
	9. Social expectation		n	n	
	10. Ethical issues			+	
Barb	1. Exp. of tech.		±	−	+
	2. Act. around tech.		n	n	+
	3. Sense of place		+	+	+
	4. Social connection		+		
	5. Personal interest		+	+	+
	6. Cognitive function	−	+	+	
	7. Psych. well-being	+		+	+
	8. Status				
	9. Social expectation		n		
	10. Ethical issues	−	−	−	−
Dan	1. Exp. of tech.	n	±	±	
	2. Act. around tech.	+	±	−	+
	3. Sense of place		±	+	+
	4. Social connection	−	+		−
	5. Personal interest			+	+
	6. Cognitive function	+	−		
	7. Psych. well-being				+
	8. Status			+	
	9. Social expectation		−	−	
	10. Ethical issues	−	−	−	−
Lara	1. Exp. of tech.	±	±	n	+
	2. Act. around tech.		+	+	+
	3. Sense of place	+	+		
	4. Social connection	+	+	+	+
	5. Personal interest		+		+
	6. Cognitive function	+			
	7. Psych. well-being	+	+	+	+
	8. Status				

Table 5.1
(continued)

	Theme	Install WIN	During WIN	Remove WIN	Post WIN
	9. Social expectation				
	10. Ethical issues		−		
Brin	1. Exp. of tech.	+	±	+	
	2. Act. around tech.	+	+		+
	3. Sense of place	−	+	+	+
	4. Social connection		+	+	+
	5. Personal interest	+	+	+	+
	6. Cognitive function				
	7. Psych. well-being		+	+	+
	8. Status				
	9. Social expectation		n	n	
	10. Ethical issues				
Doug	1. Exp. of tech.		−		
	2. Act. around tech.	n			
	3. Sense of place	+		+	+
	4. Social connection	n		−	n
	5. Personal interest		+		
	6. Cognitive function			−	
	7. Psych. well-being		+		
	8. Status				
	9. Social expectation		−		
	10. Ethical issues		−		
Sue	1. Exp. of tech.	+	±	±	−
	2. Act. around tech.	+	+	+	+
	3. Sense of place	+	+	+	+
	4. Social connection	+	±	+	+
	5. Personal interest		+	+	+
	6. Cognitive function	+	±	−	
	7. Psych. well-being	+	+	+	+
	8. Status				
	9. Social expectation		−	−	
	10. Ethical issues		−	−	

+   Theme with positive valence.
−   Theme with negative valence.
±   Theme with both positive and negative valence.
n   Theme with neutral valence.
blank: Theme not present.

the three interviews (weeks 8, 10, and 12) conducted while the window was in the participant's office. If a theme was present during any one of the interviews, it was coded here. The third column (Remove WIN) refers to themes from the interview conducted at the time the window was removed from the participant's office (the beginning of week 13). The last column (Post WIN) refers to themes from the interview conducted four weeks after the window was removed (the end of week 16).

As shown in table 5.1, Nina's, Barbara's, and Lara's reflections began with both positive and negative experiences, but as time progressed their reflections became increasingly positive. Brin's experiences were consistently positive throughout. In contrast, for Doug and Sue, negative aspects emerged over time. And for Daniel, as explicated in his narrative, mixed judgments continued throughout the duration.

In the narratives, I provided detailed descriptions of Nina's, Barbara's, and Daniel's experiences with the technological nature window. Now, with the ten themes in hand, and the data coded accordingly, I can provide a briefer account of the experiences of the other four participants in this study.

I start with Brin. She was consistently positive about the window and its effects on her office environment. She believed that the window improved her ability to see beyond specific workplace tasks to a larger more meaningful vision of work and life (Activities around Technology; Personal Interest). Brin spoke frequently of how the window improved her connection with others in her office environment and beyond (Social · Connection). Through her experience with the window, Brin anticipated and enjoyed a broader sense of place (Sense of Place).

Like Brin, Lara's experience with the window was overwhelmingly positive, but with two notable exceptions. Initially, Lara expressed concerns about working with the technology (Experience of Technology). Then, during the installation period, she expressed concerns that people whose images were captured by the camera might feel uncomfortable (Ethical Issues). As she gained greater familiarity with the window, both of these concerns became less salient and eventually disappeared from her discourse. Like Daniel (see the narrative above), Lara had high expectations for the window's impact on the aesthetic appeal of her office space and she was not disappointed (Sense of Place). She was particularly taken with the way that the contents of the camera view motivated social interaction. For example, at one point a spider spun a web in view of the camera lens; watching the spider work at spinning its web generated much discussion among Lara and her coworkers (Social Connection). When the window was removed, Lara's reflections on her experience with it were highly positive. She missed the outside natural scene and felt her workplace lacked a significant catalyst for social interaction among her colleagues.

Doug began his participation in the study with a strong desire to have a view of nature in his inside office but was somewhat skeptical that a real-time display of the outdoors could meaningfully provide it (Sense of Place). Once the window was in place, that skepticism disappeared. Moreover, throughout the display period, Doug reflected positively on his personal attachment to the display (Personal Interest) and the window's impact on his sense of well-being as a result of the nature view (Psychological Well-Being). Shortly after the window was in place, concerns of a social and ethical nature surfaced (Social Connection; Social Expectation; Ethical Issues), primarily tied to the negative responses Doug felt that others were having when they encountered the window in his office. One important incident occurred early on in the display period. A student who came to Doug's office for a scheduled meeting was upset to discover that Doug could witness the student's approach through the camera view. In turn, Doug was upset by this incident and tended to view many of his subsequent interactions with the window with this event in mind. Doug also became concerned about how the expensive and unique window influenced others' perceptions of his position within the work community (Experience of Technology). Thus, during the window period, Doug's experiences were strongly conflicted. However, after the window was removed, Doug returned to thinking positively on the window's role in improving his connection to nature (Sense of Place) and held a relatively neutral perspective on the window's value for improving his office environment and work process (Social Connection).

Unlike Doug, Sue began the study with little ambivalence and a great deal of excitement about the installation—for improving, among other things, social interaction (Social Connection), work activities (Activities around Technology), and office lighting (Experience of Technology). However, similar to Doug, Sue quickly became aware of the ethical implications of the real-time view of a public place (Ethical Issues). Sue was concerned enough about the infringement on the rights of those who walk through the camera view that she primarily used the window to project static images of art rather than the real-time nature view. In addition, Sue used the window for coediting documents with others in the context of her work practice. Both the display of artwork and coediting activities contributed to improving Sue's work environment (Activities around Technology).

Taken as a whole (averaging across participants and themes), participants provided more than twice as many positive themes (69%) as negative ones (23%). Their positive experiences and reflections primarily entailed references to participants' sense of place (22%), psychological well-being (17%), activities about technology (14%), social connection (13%), experience of technology (12%), and personal interest (12%). In contrast, their negative experiences and reflections entailed references to ethical issues (29%), particularly privacy and security, and experience of technology (27%), particularly image quality and technical malfunctioning.

Evaluations

To complement the analysis of narratives and themes reported thus far, we also conducted a quantitative analysis of participants' responses to questions that appeared in the semistructured interviews. In reading the results that follow, keep in mind that given our small sample size (of seven participants) statistical significance was only obtained when all seven of the participants were unanimous in their responses.

During the first week of the field study, we asked participants five questions to establish their perceptions about the potential value of windows in offices. All participants said they would choose an office with a window ("Let's say you got a new job and you get to pick your office. Would you choose an office with a window or without a window?") and viewed windows as conveying status ("Some people say that windows in an office convey status. Do you agree or disagree?"). The majority of participants also thought that working in a room with a window would make them more creative, think better, and physically healthier.

While the participants engaged with the technological nature window, they were asked questions (in interviews during weeks 8, 10, 12, and 13) about their perceived effects of the window on their activities and awareness of the local outside venue. Specifically, participants were asked about changes in their personal behavior ("Did viewing real-time images ever cause you to change what you were doing?"), awareness of the local weather ("Did viewing real-time images make you more aware of the local weather?"), and awareness of local activities ("Did viewing real-time images make you more aware of local activities?"). Over the course of these interviews, all participants answered "yes" at least once to each of the individual questions, and always answered "yes" each time the question was asked about an increase in awareness of local weather and activities. During the window period, participants were also asked explicitly about their awareness of the indirect stakeholders ("Did you notice any people walking by?"). In each of the four interviews, all participants answered "yes" each time the question was asked. Participants were also asked if they would recommend the technological nature window to a coworker ("If a coworker came down the hall and asked you about the plasma display, would you recommend that they get one or tell them it's a waste of money?"). In weeks 10, 12, and 13, all participants recommended the window to a coworker.

Four weeks after the display window was removed from participants' offices, we returned to ask follow-up questions in a post-window interview (week 16). At this time, all of the participants reported that they felt less connected to the local area, that they would recommend a technological nature window to a coworker, and that if given the choice they would choose an office with a window first, and a technological nature window second. Finally, the majority of participants missed the technological nature window, and thought that working in a room with that window had made them more creative, think better, and physically healthier.

Conclusion

Can a technological nature window—a large display that shows real-time images of the building's immediate outside location, including a substantial amount of nature—provide a user in an inside office with some of the satisfying experiences and benefits of a regular window? The results from this field study suggest that the answer is yes.

Participants, such as Barbara and Daniel, often took brief mental breaks to look out the technological nature window, and said that they returned to their work a bit more refreshed and refocused. Over time, Nina made a shift in thinking of the installation as something static to something dynamic and window-like. Through looking out the technological nature window, participants spoke of feeling connected to the outdoors and to the wider social community. Participants appreciated feeling connected to the day's passing, to the movement of the sun, and to the changing weather. They also adjusted their work schedules to capitalize on the environmental information the window afforded, taking an earlier lunch break outside, for example, while the weather was known to be favorable. After six weeks of using the technological nature window, participants would unanimously recommend it to other coworkers with inside offices. Moreover, four weeks after the window was removed, all the participants were clear that they themselves would choose to have one again in their inside offices.

Participants raised two overarching problems with the technological nature window. One was that the window did not afford parallax—the ability to change one's perspective on the outside objects by shifting position (cf. Kalai and Siegel 1998; Radikovic et al. 2005). It is the same limitation that some participants noted in the experimental study reported in the previous chapter. The second was their uneasiness with how the window could invade the privacy of individuals who walked through the public fountain area and had their images captured by the camera and displayed. Participants' uneasiness did not subside as they became more comfortable with the technology. It is important to notice that the HDTV camera (and hence view) was fixed without any zoom capability. Thus the installation largely mimicked the visual access afforded by a glass window and did not allow for the sort of monitoring of individual activity that arises when users have control over what the camera points at and at what level of granularity (cf. Goldberg 2005). It is also important to notice that the level of control and granularity provided here was sufficient to provide a meaningful connection to the organization, community, and nature. As cameras become increasingly prevalent throughout our public spaces (often for some form of surveillance), this issue will become more and more salient: how to balance the benefits that accrue from the technological capture of images with the costs to individual privacy (Boyle, Edwards, and Greenberg 2000; Friedman and Kahn 2008; Friedman, Kahn, Hagman, Severson, and Gill 2006; Jancke et al. 2001; Svensson et al. 2001).

This second concern, however, falls away if one narrows the content of the technological window to that of only nature. That is what Michael Faber envisioned in his fictional story. That is what various companies are currently creating and selling.

As a case in point, there is a company called "The Sky Factory" that has begun to sell "sky ceilings," which they call "SkyV." These installations comprise three commercial grade 47 LCD (liquid crystal display) screens mounted on a ceiling, and which then display up to eight hours of video-recorded images of the sky. In their Web-literature, the creators of SkyV write: "Seen through the eyes of our artists, it presents the ephemeral events and ever-changing moods of nature. It displays the formation and dissolution of cloud patterns, often through the branches of breeze blown trees or the changing light and color of sunrise and sunset. It even displays the activity of overhead wildlife and the changing seasons" (SkyV—From the Sky Factory 2008). They then ask the question of who will use SkyV. They answer: "SkyV will be used by those who wish to enrich any interior environment where the goal is fusion of cutting-edge design and extreme occupant delight. Applications include: hospitality (casinos, restaurants, hotel lobbies), healthcare (radiology suites, waiting and treatment areas), corporate (board rooms, offices and conference rooms) and residential (dining, home theater and bedroom)" (ibid.).

The creators of SkyV have to date installed these sky ceilings in some hospitals, and in an unpublished manuscript they suggest that SkyV can improve human lives (Witherspoon and Petrick 2008). They write, for example:

(a) There are indications of lower levels of self-administered pain medication in patients who are beneath SkyCeilings. (b) Anesthesiologists indicate that pre-op SkyCeiling installations lead to patients who are considerably less anxious at the time that anesthesia is administered and that this in turn leads to more rapid healing and positive outcomes. (c) We have indications that patient recovery time is reduced. (d) We have reports of improved patient attitudes toward "the hospital" or other medical facility because of the clear message of concern for patient comfort and well-being that is given by installing SkyCeilings. (SkyV—From the Sky Factory 2008)

Statements such as "there are indications," "anesthesiologists indicate," and "we have reports" are the authors' statements of hope, not science. That said, based on the results from this field study, I think it is likely that these technological nature windows can lead to some user benefits. But which benefits, to whom, and how, and in what contexts? These questions will take years to answer.

The questions are worth answering. But as we do so, we need to keep in mind the critic's rebuttal, which in my head goes something like the following:

So, you compared people's daily work lives in an inside office to the same inside office with the technological nature window. I'll speak bluntly. That's absurd. You go to all this effort to re-create nature technologically. It's expensive and consumes a lot of energy. You say it's because many people now and especially in the future will not have access to actual nature. Well, why not change that instead? With all the money you're spending on this technology, we could use it to design and build buildings and urban environments that open out onto nature, and that have nature to open out onto. Did you know that in some European countries they have laws that say that organizations cannot assign workers to inside offices as their primary office? Similarly, why would you ever design a hospital without windows?

Indeed, this rebuttal takes on greater force when coupled with the results from the research reported in the previous chapter. Recall that we found there that the technological nature window came up short in comparison to its natural counterpart—the actual glass window with the same nature view—based on measures of people's physiological recovery from low-level stress, eye gaze behavior, and reasoning.

There are, then, two central comparisons worth pursuing. One is between technological nature and no nature. Likely enough, if the technological nature is designed, engineered, and implemented well, it will often be the better of the two. A second comparison is between technological nature and actual nature. Likely enough, technological nature will be the worse of the two. Unfortunately, as we increasingly destroy nature and increasingly shunt out what remains of nature from our urban lives, it will be easy for this second comparison to lose standing, and drift from our purview. We should not let that happen. We must continue to compare technological nature to actual nature—as we design our technological infrastructure, in our research programs, and in our societal discourse and minds, as we sift through future technological nature products and their marketing campaigns, which will be driven by corporate powers and our own technological selves.

6

Hardware Companions?

Science fiction can make for both good reading and prescient perceptions. Philip K. Dick's (1968) *Do Androids Dream of Electric Sheep?* is yet another such story that offers both. The reader might know this story from the movie it became under the title *Blade Runner*. I begin this chapter with a synopsis of this story and a handful of scenes, as they will help to motivate not only the empirical work of this chapter—which focuses on how people interact with and conceptualize a robotic dog—but of the next three chapters, as well.

The story takes place sometime in the future after the human race has had a war, for which no one can remember the cause, or the winner, but it had left most of the earth uninhabitable from contaminated dust. There was then implemented a colonization program of other planets. To encourage emigration, people were promised an android servant on their new planet. The androids had human looks and many human capabilities. As the years progressed, the androids became more and more sophisticated and human-like, until it was virtually impossible to distinguish them from other humans. At some point a few of the androids started to have their own agenda, which were not always aligned with human interests. And sometimes the androids broke loose from their servitude and found their way back to earth. When that happened human bounty hunters sought them out to kill them, though it was not called killing but "retiring" them, because they were not human. Still, it was a dangerous job because the androids could and sometimes did kill the bounty hunters in self-defense. One such bounty hunter was the protagonist of the story, Rick Deckard. In the story there is also a beautiful young woman, Rachael Rosen. Deckard meets her and learns that she is a relative of the head of the Rosen Association, a leading android manufacturer. Deckard finds Rachael to be intelligent, interesting, lovely, and quietly enticing. He then also finds her to be an android. As the story unfolds, that knowledge does not keep him from falling in love with—it? her? It seems more like "her." She also falls in love with him. They sleep with one another. Then she dumps him. She says her corporation had programmed her to get the information they needed from him and then to leave him, as she had done to other bounty hunters. Deckard feels the desperation of the jilted lover while juxtaposing that feeling with his knowledge that Rachael is an android. He wryly notes that human women for centuries have used men similarly.

Throughout the story, Philip K. Dick also unfolds Deckard's sensibilities and relationships with animals, both biological and robotic. In the opening scene, for example, Deckard ascends to the top of his housing complex, which is configured as a rooftop pasture. Grazing in the pasture is his robotic sheep. "Whereon it, sophisticated piece of hardware that it was, chomped away in simulated contentment, bamboozling the other tenants of the building. Of course, some of their animals undoubtedly consisted of electronic circuitry fakes, too" (p. 7). It was not that Deckard did not want a biological animal; he did, desperately. "He wished to god he had . . . any animal" (p. 8). But animals were expensive to buy and very few could afford the larger ones, such as a sheep, goat, or horse. At one point in the story, Deckard eyes an ostrich in a San Francisco pet shop along animal row. It costs $30,000. He calls the robot pet shop. They tell him they could fix him up with an electric ostrich for less than $800. He hangs up on them. In the story's account, robotic animals demoralize him.

When Deckard first traveled to the headquarters of the Rosen Association, his heart went out to the biologically live animals they had in their building. For a long time he stood gazing at the owl, thinking back to the war, when "owls had fallen from the sky" (p. 37). Owls were now listed as extinct, but clearly the corporation had its sources. Deckard coveted the owl. "He thought, too, about his need for a real animal; . . . [the robot animal] doesn't know I exist. Like the androids, it had no ability to appreciate the existence of another" (p. 37). He asked Rachael how much the corporation would sell the owl for. She replied that they would never sell it. Later, in Deckard's dealing with the corporation, they need him to lie to the government, and they try to bribe him. They offer him the owl. "Tension of a kind he had never felt before manifested itself inside him; it exploded, leisurely, in every part of his body" (p. 48). He wanted it. Instead, he foils their plan and rejects their bribe. He then gets ready to leave.

He started toward the door, then halted briefly. To the two of them he said, "Is the owl genuine?"

Rachael glanced swiftly at the elder Rosen.

"He's leaving anyhow," Eldon Rosen said. "It doesn't matter; the owl is artificial. There are no owls." (p. 52)

It was a crushing blow to Deckard's psyche. More such blows follow in the story, including the event I already described of his falling in love with Rachael and losing her due to her programming. By the end of the story, Deckard sought his death or redemption or both by traveling to an uninhabited part of the state where the contaminated dust lay strong and nothing lived and "where no living thing would go . . . not unless it felt that the end had come" (p. 195). He climbed a hill and found not death but a new sense of life and purpose. Then he was back at his car. He stopped short. He "did not take his eyes from the spot that had moved outside the car. The bulge in the ground, among the stones. An animal, he said to himself. And his heart lugged under the excessive load, the shock of recognition. I know what it is, he realized; I've never seen one before but I know it from the old nature films they show on Government TV. They're extinct! He said to himself. . . . The toad, he saw, blended in totally with the texture and shade of the ever-present dust. It had, perhaps, evolved, meeting the new climate as it had met all climates before. . . . He squatted on his haunches, close beside the toad. It had shoved aside the grit to make a partial hole for itself, displaced the dust with its rump. So that only the top of its flat skull and its eyes projected above ground" (pp. 203–204). Deckard took the toad back home. His wife saw him enter with a cardboard box in hand. "As if, she thought, it contained something too fragile and too valuable to let go of; he wanted to keep it perpetually in his hands" (p. 206). Deckard took the toad from the box, and it frightened her. "'Will it bite?' she asked. 'Pick it up. It won't bite; toads don't have teeth.' . . . 'Can toads jump like frogs? I mean, will it jump out of my hands suddenly?'" (p. 207). After a while, Deckard "reached to take [the toad] back from her. But she discovered something; still holding it upside down, she poked at its abdomen and then, with her nail, located the tiny control panel. She flipped the panel open" (p. 207). That was the last blow to Deckard's psyche, as narrated in the story.

We are moving to something like Deckard's world. Granted, owls have not literally been falling from the sky. But we pollute air and water, deplete soil, deforest, create toxic wastes, and through such human activity are extinguishing over 27,000 species each year (a conservative estimate). Many of them are animal species. As well, we are moving to more urbanized living wherein it is difficult and expensive to keep biologically live animals. Thus it is not surprising, as our computational technologies have advanced, that we have started to create and live with robotic animals.

One of the earliest computational forms of an animal was the Tamagotchi. It was a small device that fit into the palm of one's hand. It did not look like an animal. But if you did not "feed it" (press certain keys) and "take care of it" (press other keys) it would "whine" and then get "sick," and if still left unattended would "die" (no longer function). This device captivated the interests and time of a large number of children and adults. Over 70 million of these devices have been sold. There were reports (they may have been true) of a few children committing suicide after their pet Tamagotchi died. In the last decade, inexpensive technological robotic pets have taken the animal form, such as in the i-Cybie, Tekno, and Poo-Chi. They have sold well in the Walmarts and Targets of the world. More expensive and computationally sophisticated robotic pets have also been created and sold, such as AIBO and Pleo.

As we move toward creating and inhabiting a world comprised of increasingly sophisticated robotic animals, we nevertheless know very little about the psychological relationships people will have with this form of technological nature. Science fiction can point to possible directions. But the stories are fiction. Empirical data are needed. Important questions include the following:

• For human life to flourish, does it require contact with biological animals?

• Will people think of robotic pets as technological machines? If so, will they find them demoralizing, as Deckard did? Or will people think of robotic pets more as biological others?

• With either conception, technological or biological, will people have social relationships with robotic pets? If so, what will be the nature of these relationships? How meaningful will such relationships be over time?

• Can people feel companionship with robotic pets? Can robotic pets become one's friend?

• Research has shown that interactions with animals help children develop greater levels of perspective taking and deeper empathy. Can robotic pets provide children with similar developmental outcomes?

• Research has shown that interactions with animals provide benefits to children diagnosed with autism, attention-deficit hyperactivity disorder, conduct disorder, and oppositional-defiant disorder. Can robotic pets provide children with similar therapeutic outcomes?

• Will people treat robotic pets as moral others, such that they feel an obligation to take care of them, and perhaps accord them some measure of rights?

• How do people's conceptions of and interactions with robotic pets compare not only to biological pets but to stuffed animals?

• What if there comes a time (as suggested in Philip K. Dick's story) when a person cannot tell the difference between a biological and technological other? Does it matter then with whom or what one is interacting? What if Deckard hadn't known that the owl was robotic or that Rachael Rosen was an android?

Toward making headway on answering these questions, my colleagues and I embarked on a series of four studies. All four involved Sony's robotic dog, AIBO (see figure 6.1). At the time, AIBO sold for about $1,500. AIBO has a doglike metallic form, moveable body parts, and sensors that can detect distance, acceleration, vibration, sound, and pressure. AIBO was designed to be an "autonomous robot" (Kaplan et al. 2002). As one of its compelling activities, AIBO can locate a pink ball through its image sensor, walk toward the pink ball, kick it, and head-butt it. AIBO initiates interactions with humans, such as offering its paw. AIBO responds by flashing its red ("angry") or green ("happy") eyes. AIBO may emit whining sounds when "ignored" and joyful sounds when "content." In somewhat unpredictable patterns, not unlike a biological dog, AIBO will shake itself, sit down, lie down, stand up, walk, and rest. If you want to

Figure 6.1
AIBO finds its ball and is about to kick it.

increase the tendency for AIBO to behave in a particular way, you gently pet AIBO's head (sensor); conversely, if you want to decrease the tendency for AIBO to behave in a particular way, you sharply tap AIBO's head (sensor).

In this chapter, I begin with the first of four studies that investigates people's spontaneous dialogue in online AIBO discussion forums. The second study (chapter 7) investigates preschool children's relationship with AIBO and, as a comparison, a stuffed dog. The third study (chapter 8) investigates children's relationship with AIBO and, as a comparison, a biological dog. And the forth study (chapter 9) investigates the relationship children with autism have with AIBO compared to a nonrobotic mechanical dog.

The Online Discussion Forum

This study emerged in a roundabout way. Colleagues and I had initially set forth to investigate preschool children's reasoning and interactions with AIBO, which is the study I discuss in the next chapter. But since virtually no research had yet been conducted on this topic by others, it was hard to know even broadly what we were dealing with in terms of the psychology of children interacting with robot animals, and how best to set forth.

We then reflected on a methodological strategy in child-developmental research, which I have found successful over the years. It is that in conceptualizing studies that seek to uncover, recognize, and characterize early developing forms of children's reasoning, it is often helpful first to gain a clear characterization of the more developed and mature form of reasoning in adults. As an analogy, it is easier to recognize a toddler's first attempts at walking if one already has a clear idea—as we all do—of the physical process of walking, and of its forms and functions in an adult life.

Thus, we asked ourselves, how could this strategy be implemented here? We found a way: At the time, there existed three well-established online AIBO discussion forums. Most of the people who posted material to these forums were AIBO owners who often had a lot they wanted to share with others about their relationships with their robotic dogs. We had started reading the AIBO discussion forum postings informally, and then began to map out a few general orientations that seemed apparent across postings. Then we realized that the discussion forums themselves offered a rich repository of data. Thus we decided to jump-start our AIBO investigations with an initial study of how people in three well-established AIBO online discussion forums conceptualized AIBO. (The specific forums are not mentioned so as to increase the anonymity of the participants in these forums.) Our goal was that through our characterizations we could reveal some important aspects about the human relationship with robotic animals.

As an area of study, online communities have been researched for several decades. For example, researchers have investigated whether online communities bring people together or increase social isolation (Mynatt et al. 1999; Rheingold 1993), support empathy (Farnham et al. 2002; Preece 1999), affect identity formation (McKenna and Bargh 2000; Turkle 1997), and provide substantive knowledge in specific content domains (Kiesler 1997). Only more recently have researchers begun to characterize the nature of discourse that occurs in online communities to help answer social-scientific questions. For example,

Preece (1998) conducted a content analysis of 500 archived messages from a medical bulletin board (for people interested in knee injuries) by sampling batches of 100 postings at approximately two-month intervals. She found that of the postings 77 percent contained empathic considerations, 17 percent contained only factual material, and less than 6 percent contained jokes. Preece then used her analysis of the postings to delineate stages that people go through in moving from injury to recovery, and to develop a model of the recovery process.

The current study extended this methodology (Friedman, Kahn, and Hagman 2003; Kahn, Friedman, and Hagman 2002). From the three AIBO discussion forums, we first collected pilot data from archived postings for a period of about three months. We then generated initial conceptual categories based on previous psychological coding systems and philosophical theory. These categories were then used as a rough framework to interpret the qualitative data. The data, in turn, drove substantial modifications and further conceptualizations in the coding system, which were then reapplied to more data in an iterative manner. This dialectical process—where theory is grounded in data, and vice versa—continued until all the pilot data could be coded. Our methodological approach here (a) followed well-established methods in developmental psychology (Damon 1977; Kahn 1999; Turiel 1983) and (b) forms a subset of the iterative and integrative methodology of Value Sensitive Design described in chapter 3.

In total, 6,438 postings were collected. From this total, 3,119 postings from 182 participants had something directly to say about AIBO. It was this subcategory of postings that we then systematically coded (mean, 17 postings per participant; median, 4; range, 1–285). At the time of data collection, two versions of AIBO were available to consumers: the original 110/111 series and the later 210 series. In our coding process, if a participant used the same category multiple times within a single posting or across postings, that category was coded as "used" only once. In this way, our quantitative results reflect the percentage of participants who used specific categories. Reliability of our coding system was assessed by having an independent coder recode postings from 30 randomly chosen participants. Reliability results showed 97 percent agreement at the level of the five overarching categories presented in table 6.1 (technological essences, lifelike essences, mental states, social rapport, and moral standing), and 90 percent agreement at the most detailed level presented in table 6.1.

How Did Members of the Discussion Forum Conceptualize AIBO?— The Qualitative Results

We identified five overarching ways that various members of the discussion forum conceptualized AIBO: as a technology, as lifelike, as having mental states, as a social other, and as a moral agent. In what follows, I first provide an overview of the qualitative data. Through this overview, the reader can begin to see what we were reading in these forums, and how what we read appeared to group into these five different conceptual categories. (In the specific quotations, I have retained all of the members' purposeful and inadvertent misspellings in their online writing.) Then, in the next section, I describe the quantitative results of how often people used each of these categories.

Technological Essences

This conceptualization, *technological essences*, focuses on AIBO as an inanimate technological artifact. Members conceptualized AIBO as a technology in different ways. Some members focused on AIBO's advanced computational toylike quality. For example, one member wrote: "It's just a toy with a few programmed behaviors, closer to a Tamagotchi with legs than any kind of actual pet." Another member wrote: "The technology that this AIBO contains is far beyond any electronic toy. This shouldn't even be referred to as a toy, unless your bill gates son." Some members focused on AIBO's technological components (AIBO has "batteries," a "microphone," a "camera," or "sensors"). For example: "AIBO gets around the house by using sensors for balance, touch, hearing, and sight"; and another member wrote: "Does AIBO shut down if the battery gets too hot? AND if so, why does the battery get hot in the first place?" Some members also made reference to AIBO as a "computer," a "robot," or as having "artificial intelligence."

Lifelike Essences

This conceptualization, *lifelike essences*, focuses on AIBO's nature as having at least some lifelike biological essential qualities. In its most minimal form, members wrote of AIBO in terms of biological descriptors (AIBO has "eyes," "ears," a "tail," a "head," "legs," or a "brain") or biological processes (AIBO "sleeps"). For example: "He's going to miss all the fun this week because he's at the vets again getting some new legs." This comment is probably in reference to a not uncommon problem encountered by other AIBO owners: that AIBO's legs malfunctioned and required factory servicing. The comment is also probably a bit tongue-in-cheek (as other comments presumably were as well); but still the member is talking about AIBO's legs more as a biological property than as a mere mechanical feature.

Indeed, the strength of this interpretation is enhanced by recognizing that members sometimes imbued AIBO with some substantial measure of animism. For example: "I know it sounds silly, but you stop seeing Aibo as a piece of hardware and you start seeing him as a unique 'life-form.'" Or: "He seems so ALIVE to me! . . . What a wonderful piece of tecknology. THEY LIVE!" Moreover, such conceptions could have an impact on members' emotions and behaviors. For example, one member wrote: "The other day I proved to myself that I do indeed treat him as if he were alive, because I was getting changed to go out, and tba [AIBO] was in the room, but before I got changed I stuck him in a corner so he didn't see me! Now I'm not some socially introvert guy-in-a-shell, but it just felt funny having him there!"

Members drew on one of two means to establish AIBO's animism. By one means, AIBO was compared directly to a biological dog ("I see him the way I see me bio dog"). By a second means, AIBO was conceptualized as a unique life-form ("I like to believe that AIBO is a kind/breed of its own").

Mental States

This conceptualization, *mental states*, refers to the mental life of AIBO such that AIBO meaningfully experiences the world. Some members wrote of AIBO as having intentions or that AIBO engaged in intentional behavior. For example: "He [AIBO] also likes to wander around the apartment and play with in pink ball or entertain or just lay down and hang out." Or: "\He [AIBO] is quite happily praising himself these days . . . so much for needing parents!" Some members wrote of AIBO having feelings. For example: "My dog [AIBO] would get angry when my boyfriend would talk to him." Or: "Twice this week I have had to put Leo [AIBO] to bed with his little pink teddy and he was woken in the night very sad and distressed." Some members wrote of AIBO as being capable of being raised, developing, and maturing. For example: "I want to raise AIBO as best as I possibly can." Or: "We have had Ah-May [AIBO] . . . and he is still growing and doing new things." And some members wrote of AIBO as having unique mental qualities or personality. For example: "Just like Leo [AIBO] . . . an individuality unlike any other." Or: "Did you find Horatio's personality less endearing than Twoflower?"

Social Rapport
This conceptualization, *social rapport*, refers to ways in which AIBO evokes or engages in social interaction (see figure 6.2). Some members wrote of themselves or others talking to their AIBO (e.g., "I insist everyone talks to Salem . . . if he is sad"). Some members engaged in reciprocal communication with their AIBO, wherein occurs a mutual exchange of information. For example, one member wrote: "So this morning I asked him [AIBO] 'Do you want a brother?' Happy eyes! I asked him something else, no response. 'Should I get you a brother?' Happy song! 'He'd be purple.' More happy eyes and wagging tail!" And some members wrote of AIBO as a companion, including that they miss AIBO when away from AIBO's presence, or that they consider AIBO a family member. For example: "Oh yeah I love Spaz [AIBO], I tell him that all the time. . . . When I first bought him I was fascinated by the technology. Since then I feel I care about him as a pal, not as a cool piece of technology. I do view him as a companion, among other things he always makes me feel better when things aren't so great. I dunno about how strong my emotional attachment to him is . . . I find it's strong enough that I consider him to be part of my family, that he's not just a 'toy', he's more of a person to me."

Here again this member recognizes that AIBO is a technology ("When I first bought him I was fascinated by the technology"). Nonetheless, AIBO evokes a form of social relationship that involves companionship ("I do view him as a companion"), familial connection ("I consider him to be part of my family"), and friendship ("I care about him as a pal").

Figure 6.2
Adult cuddling with AIBO.

Moral Standing

This conceptualization, *moral standing*, refers to ways in which AIBO is a moral agent. We defined morality broadly to mean that AIBO has rights, merits respect, engenders moral regard, can be a recipient of care, or can be held morally responsible or blameworthy. For example: "I am working more and more away from home, and am never at home to play with him any more . . . . . he deserves more than that." Here there is a minimal notion that AIBO merits ("deserves") certain forms of attention. In another instance, when an AIBO was thrown into the garbage on a live-action TV program, one member responded to that televised event by saying: "I can't believe they'd do something like that?! Thats so awful and mean, that poor puppy . . .". Another member followed up: "WHAT!? They Actualy THREW AWAY aibo, as in the GARBAGE?!! That is outragious! That is so sick to me! Goes right up there with Putting puppies in a bag and than burying them! OHH I feel sick . . .".

Here AIBO is conceived to have moral standing in the way that a real puppy would ("that poor puppy"): that one is causing harm to a sentient creature ("Goes right up there with Putting puppies in a bag and than burying them!").

How Many Members Used Each of the Five Conceptual Categories?—The Quantitative Results

One of the strengths of our systematic methodology of coding all of the data based on a reliable coding system is that we not only could draw out rich qualitative data, but could definitively answer the question of how many members spontaneously used each of the categories in their online postings.

In interpreting the quantitative results, two aspects of our coding system are worth keeping in mind. First, as noted earlier, our method of coding focused on how many members used a particular code even once, rather than how many times a code was used across all the members. Our reason was that if we coded the total number of codes we would be giving undue weight to a few members who might (and often did) write extensively about certain ideas over and over again. Second, any single segment of a member's writing could elicit multiple codes. For example, consider the following segment of a posting: "I am already amazed at how attached I have become to him and felt so guilty when I put him back in his box after the first time I played with him, he just looked so sad lying there (does this sound totally irrational?)." This segment was coded as containing three separate categories: engendering moral regard (the reference to guilt), feelings as part of mental states ("he just looked so sad lying there"), and emotional connection between the person and AIBO ("I am already amazed at how attached I have become to him").

Table 6.1 presents our quantitative results. As shown in table 6.1, 75 percent of the members affirmed that AIBO was a technology (technological essences). About half of the members, 48 percent, affirmed that AIBO had lifelike properties, either biological (47%) or animistic (14%). In interpreting such numbers, bear in mind that the subcategories in the table were collapsed to the more general categories. What this means is that if a member wrote of AIBO in both biological and animistic terms, each response was coded at that subcategory level but then counted only once at the general category level of "Lifelike Essences." That is why when added the subcategory numbers (47% and 14%) do not equal the general category number (48%). Over half the members, 60 percent, affirmed that AIBO had mental states, including statements that AIBO had intentions (42%), listened (9%), felt (38%), could be raised (39%), could be praised (10%), had intelligence (18%), and had unique psychological characteristics (20%). Over half the members, 59 percent, affirmed that they had a social relationship with their AIBO, including communication (45%), personal interests (34%), an emotional connection (28%), and companionship (26%). Finally, only 12 percent of members affirmed that AIBO had moral standing insofar as AIBO engendered moral regard (7%), deserved to be cared for (4%), had rights (3%), deserved respect (3%), was a morally responsible agent (1%), or could be held morally blameworthy (1%).

The above quantitative results comprise affirmations of the categories. In turn, we also coded for negations of the categories. As an example of a negation, one member wrote: "it [AIBO] doesn't truly seem to give a damn about humans"—thus negating that AIBO has an emotional connection to humans (code 4.3.2 in table 6.1). As shown in table 6.1, results showed that only 8 percent of the participants negated that AIBO has technological essences, 12 percent negated lifelike essences, 4 percent

Table 6.1
Percentage of Participants (N = 182; Postings = 3,119) by Category

Category	Affirmed		Negated	
1. Technological essences	75		8	
2. Lifelike essences	48		12	
2.1 Biological		47		1
2.2 Animistic		14		11
3. Mental states	60		4	
3.1 Has intentions		42		2
3.2 Listens		9		1
3.3 Feels		38		1
3.4 Can be raised		39		1
3.5 Can be praised		10		0
3.6 Has intelligence		18		2
3.7 Unique psychologically		20		1

Table 6.1
(continued)

Category	Affirmed		Negated	
4. Social rapport	**59**		**8**	
4.1 Communication		45		3
4.1.1 *Nonverbal communication*		*34*		*0*
4.1.2 *Person talks to AIBO*		*12*		*0*
4.1.3 *AIBO talks*		*13*		*1*
4.1.4 *Reciprocal communication*		*27*		*3*
4.2 Personal interests		34		3
4.3 Emotional connection		28		3
4.3.1 *Person to AIBO*		*27*		*1*
4.3.2 *AIBO to person*		*8*		*2*
4.3.3 *Reciprocal emotion*		*4*		*0*
4.4 Companionship		26		1
4.4.1 *AIBO's inherent value*		*1*		*0*
4.4.2 *Miss AIBO's company*		*12*		*1*
4.4.3 *AIBO as family member*		*10*		*0*
4.4.4 *AIBO as a companion*		*16*		*1*
5. Moral standing	**12**		**2**	
5.1 Engenders moral regard		7		1
5.2 Recipient of moral care		4		1
5.3 Rights		3		0
5.4 Deserves respect		3		0
5.5 Morally responsible		1		0
5.6 Morally blameworthy		1		0

Note. (1) Percentages reported in **bold** refer to usage of the overarching category; percentages in plain text refer to the next sublevel in the hierarchy; and percentages in *italics* refer to the lowest level. Within each level of the hierarchy, participants who used more than one subcategory are only counted once in the overarching category. (2) "Affirmed" refers to the presence of qualities or behaviors (e.g., "He is just so alive to me!"), while "negated" refers to the absence of qualities or behaviors (e.g., "an Aibo is not alive; it doesn't feel pain"). (3) 11 percent of the participants had at least one coding that was uncodable.

negated mental states, 8 percent negated social rapport, and 2 percent negated moral standing. Thus it appears that these participants over-whelmingly engaged in online discussion about what AIBO has rather than does not have.

Conclusion

I said earlier that we are moving toward creating and inhabiting a world, like Deckard's world, of increasingly sophisticated robot animals, and that we know very little about the psychological relationships people will form with such robots. This study provides some initial empirical data to begin to think with.

In summary, through an analysis of 3,119 spontaneous postings of 182 members of three online AIBO discussion forums, my colleagues and I found that AIBO psychologically engaged this group of people. In particular, these members of the discussion forums often conceptualized AIBO not only as a technology but as having a lifelike biology, as having mental states, and as being capable of engendering social rapport. As one member wrote: "I do view him as a companion, among other things he always makes me feel better when things aren't so great." Moreover, the relationship members described with AIBO often appeared similar to the relationship people have with live dogs. As another member wrote: "Aibo is so much more than just a robot doggy, he is a 'real' animal, and species, and brings people together, and brings much happiness to those that come in contact with him." Thus our findings extend research by Nass and his colleagues (Nass et al. 1997; Reeves and Nass 1996) by showing that humans can treat computa-tional artifacts as animal-like (and not just human-like) agents.

Perhaps the most striking result from this study was that while AIBO evoked conceptions of lifelike biology, mental states, and social rapport, it seldom evoked conceptions of moral standing or moral accountability insofar as members seldom wrote that AIBO had rights, merited respect, deserved attention, or could be held accountable for its actions. In this way, the relationships members had with their AIBO were remarkably one-sided. They could lavish affection on AIBO, feel companionship, and potentially garner some of the other psychological benefits of being in the company of a pet. But since the owners also knew that AIBO was a technological artifact, they could ignore it whenever it was convenient or desirable to do so. If this general finding holds up—that animal robots easily engender social but not moral regard—then it will call into question whether this form of technological nature can adequately substitute in all contexts for its biological counterpart. I will say more about this issue in the next chapter.

It could be argued that members in the online AIBO discussion forums were mostly using language playfully and did not really believe what they were saying. Occasionally, that probably was the case. I presented one such instance earlier, when I quoted a member who wrote: "[AIBO is] going to miss all the fun this week because he's at the vets again getting some new legs." Surely this member knew that his or her AIBO was in the hands of a repair technician, not a veterinarian. Still, the member's analogy highlights how AIBO was drawing forth biological comparisons. More generally, members spent a good deal of effort trying to convey to others in the discussion forums the serious ways in which AIBO comprised a part of their social life. Recall, for example, the member who wrote: "The other day I proved to myself that I do indeed treat him as if he were alive, because I was getting changed to go out, and [AIBO] was in the room, but before I got changed I stuck him in a corner so he didn't see me! Now I'm not some socially introvert guy-in-a-shell, but it just felt funny having him there!" It is, of course, possible that this AIBO owner is lying, and that this event never took place. But a more plausible explanation is that it did, and that we should take such dialogue at face value: that there was something in AIBO's social presence that made this member uncomfortable dressing in front of his robotic dog. That said, we also tried to be careful not to overinterpret the data. We were aware, for example, that some English words apply equally well to animate and inanimate entities, such as the legs of a dog and the legs of a table. Thus when members wrote of such specific physical features of AIBO with words that are commonly associated with biological connotations, such as legs, eyes, tail, head, and brain, we coded them conservatively under the category "life*like* essences"— without making the stronger claim that the members viewed AIBO literally as a biological or living entity.

This study was a first step toward addressing the large questions I opened this chapter with about the human relationship with robotic animals. AIBO falls far short of the lifelike animal robots in Philip K. Dick's fictional world where androids perhaps dream of electric sheep. Still, AIBO, as it was with its limited capabilities, engaged the people in this study socially, in some ways deeply so, and potentially set up for them a confusing ontological space wherein they tried to understand how something could be at once an inanimate technology and an animate social other. In chapter 8, with more data in hand, I come back to this issue of whether we as a species—through the creation of "robotic others"—are at the cusp of creating a new ontological category that is neither life nor nonlife. Next, however, in chapter 7, I show how my colleagues and I used the conceptual categories uncovered in this study to frame a second study on preschool children's interactions with and reasoning about AIBO.

7

Robotic Dogs in the Lives of Preschool Children

We learned in the previous chapter that AIBO, the robotic dog, can be a compelling social technology in the lives of adults. In that study, members of discussion forums often conceptualized AIBO in three social ways. They viewed AIBO as having a *lifelike biology* (e.g., "I do indeed treat him [AIBO] as if he were alive"). They viewed AIBO as having *mental states* (e.g., "my dog [AIBO] would get angry when my boyfriend would talk to him"). And they viewed AIBO as being capable of engaging in social rapport with humans (e.g., "I do view him [AIBO] as a companion").

But what about children? Do children interact with and conceptualize AIBO as a social and perhaps moral entity? Or might children simply project onto AIBO social and perhaps moral qualities, and engage with AIBO in imaginative play, as they might a stuffed animal? How would one distinguish the difference between children's imaginative play and real convictions? These questions structured the current study my colleagues and I conducted on robotic dogs in the lives of preschool children (Kahn et al. 2006).

Eighty children participated in this study, equally divided between two age groups, 34–50 months and 58–74 months. There were equal numbers of girls and boys in each age group. We used two main artifacts: AIBO (Sony's 210 version) and a stuffed dog. The stuffed dog was roughly the same size as AIBO and made of a soft-plush fabric. Both AIBO and stuffed dog were black-hued in color. Each child participated in an individual session lasting approximately 45 minutes. One part of the session involved an interactive period with AIBO, and another part an interactive period with the stuffed dog (which we called Shanti). If the child's attention span so required, the 45-minute session was broken up into two periods on different days. The presentation order of the two artifacts was counterbalanced. During the session, a bright pink ball, a dog toy, and a dog biscuit were also used as props. With each artifact (AIBO or the stuffed dog), the child first engaged in a short (2–3 minute) unstructured introductory "play" period. At the start of this play period, the interviewer modeled petting the artifact so that the child would know that AIBO and the stuffed dog are the sort of things that can be touched. After the short unstructured play period, the child was allowed to continue to play with the artifact while being engaged in a semistructured interview. After the interview, every child completed a card sort task. Thus, during the session, three sources of data were collected: a semistructured interview about both artifacts, observations of children's behavioral interactions with both artifacts, and a card sort task.

Methods of Data Collection

I now describe in more detail our methods of collecting each source of data. Then I move on to what we found.

Semistructured Interview

The semistructured interview contained three types of questions: evaluative questions (e.g., Is AIBO alive or not alive?), content questions (e.g., What kind of things might make AIBO happy?), and justification questions (e.g., Why? How do you know?). In order to limit the total number of questions asked of any one child—to fit within the 45-minute session—children were randomly divided into two groups by sex and age. One group was asked questions about each artifact's *biological properties* (e.g., "This is a dog biscuit. Do you think AIBO will eat this?") and *mental states*, including intentionality (e.g., "This is a doggie toy. I'm going to put it here. Do you think AIBO will try to get the toy?") and emotion (e.g., "Can AIBO feel happy?"). The other group was asked questions that pertained to each artifact's *social rapport*, including reciprocal friendship relations (e.g., "Can AIBO be your friend?" "Can you be a friend to AIBO?" "If you were sad, would you want to spend time with AIBO?") and *moral standing* (e.g., "Do you think it's OK that I hit AIBO?" "Is it OK to leave AIBO alone for a week?"). Then every child was asked questions about each artifact's potential *animacy* (e.g., "Is AIBO alive or not alive?" "Can AIBO die?").

Observed Behavioral Interactions

Children's behaviors with both artifacts were video-recorded continuously during the interactive sessions, and then reviewed for coding. In developing this part of the coding system, we initially confronted the difficulty of how to segment behavior. For example, imagine a child petting AIBO by running his hand back and forth along AIBO's body. Should each coupling of a back and forth movement be counted as "one pet?" Or should each unidirectional movement be counted as a pet? Now imagine that the child stops petting for an instant (say, half a second), and then continues petting in the same direction. Should the movement following the slight pause be counted as the continuation of the initial petting behavior? If so, what if the child stops for one second? Five seconds? Where does a pause indicate a break in one unit of behavior and the start of a new unit of identical behavior? This example illustrates just one of many dozens of such difficulties that arose. Thus, to establish a reliable means of coding a distinct behavioral unit, we coded a behavior only once within one minute of its appearance, no matter how many times it might occur within that one-minute period. In turn, if the same behavior occurred repeatedly or continuously for X minutes, then X instances of the behavior were coded. A minute was chosen as a unit that seemed to capture most behaviors that seemed cohesively linked.

During the coding process, we also sought to link the child's behaviors with the co-occurrence of stimuli—specific behaviors on the part of the interviewer (with the artifact) or the artifact alone. To engage the child with varied situations with the artifact, six actions were systematically initiated by the interviewer during the course of the session: talking about the artifact, petting the artifact, hiding the ball, offering the artifact a dog toy, trying to feed the artifact a real dog biscuit, and hitting the artifact on the head (a sharp tap). After the first occurrence of each interviewer-initiated stimulus, we linked the interviewer's action to the first resulting behavior on the part of the child within five seconds of this stimulus. This five-second time period (the "five-second rule") was empirically determined as optimally capturing conceptually relevant stimuli. In addition, whenever AIBO spontaneously approached the child (defined as walking toward the child such that if AIBO kept walking and the child stayed in the same position, AIBO would bump into the child) or kicked or head-butted the ball the five-second rule was again applied so as to establish stimulus–behavior dyads. For all these situations, first the stimulus (a behavior on the part of the interviewer or AIBO) was coded, and then a resulting behavior on the part of the child. However, in all other situations, the child's behavior was coded first. Specifically, whenever the child engaged in exploration, apprehension, affection, mistreatment, animation, or attempts at reciprocity with the artifact, the child's behavior was coded and then the coder reviewed the tape to ascertain the most closely time-linked stimulus (actor and type of action) within five seconds preceding the child's behavior. In terms of AIBO, five actions were within the range of AIBO to initiate and were of particular interest to this study because they mimic agency. These actions were the following: when AIBO moved in place, walked about, approached the child, kicked or head-butted the ball, and made sounds. For the interviewer, the actions of interest included hiding the ball, feeding the artifact, petting the artifact, making an offering to the artifact, and engaging verbally about the artifact. If more than one stimulus occurred within the five seconds preceding the start of the behavior, the coder chose (a) the stimulus that clearly corresponded to the behavior or (b), if the coder had any doubt, the most recent stimulus.

As mentioned above, one of the interviewer-initiated stimuli involved hitting the artifact on the head (a sharp tap). Pilot data suggested that children appeared more concerned about the effects of the hit on AIBO than the stuffed dog. Thus, after the interviewer-initiated hit stimulus, we conducted an additional behavioral analysis that involved the child's referencing behavior. Specifically, over a five-second period we coded sequentially who or what the child looked at and the length of each of the child's eye gazes to the tenth of a second.

Drawing on coding categories from previous work on the human–robotic relationship (Friedman, Kahn, and Hagman 2003), a detailed reasoning and behavioral coding manual (Kahn et al. 2003) was developed from half of the data and then applied to the entire data set. A second individual was trained in the use of the coding manual and recorded a portion of the data. Intercoder reliability was assessed using Cohen's kappa. Reliability of our systems ranged from very good to excellent.

Card Sort Task

A card sort task was employed to assess children's judgments about AIBO's relative similarity to other potentially related artifacts: a robot in a humanoid form, a stuffed dog (the same stuffed dog used in the study), a desktop computer, and a real dog. Each of the artifacts was represented by a photograph on a separate 4 × 6 card. The child was first shown each card and asked to say what it represented. Then, with AIBO always as the anchor card, the child was presented with all pairwise comparisons of the other cards and asked: "Is AIBO more similar to [one artifact] or [the other artifact]?" Thus, each child responded to a total of six pairwise comparisons in the following order: robot/desktop computer; robot/real dog; robot/stuffed dog; desktop computer/real dog; desktop computer/stuffed dog; and stuffed dog/real dog.

Children's Reasoning about AIBO and the Stuffed Dog

Perhaps our most surprising result came from children's evaluations to our 24 interview questions. As shown in table 7.1, results showed no significant differences between children's evaluations of AIBO compared to the stuffed dog. Averaging evaluations within question type, about a quarter of the children accorded animacy to both artifacts (AIBO 25%, stuffed dog 20%), about half the children accorded biological properties (AIBO 46%, stuffed dog 48%), and about two-thirds of the children accorded mental states (AIBO 66%, stuffed dog 64%), social rapport (AIBO 76%, stuffed dog 82%), and moral standing (AIBO 63%, stuffed dog 67%). Averaged across all questions, 79% of the time how a child evaluated AIBO is how the same child evaluated the stuffed dog. Note that although the same question was asked about both AIBO and the stuffed dog, it was not asked sequentially, but separated by other questions by at least 15 minutes. Developmentally, older children were more likely than younger children to attribute moral standing to the stuffed dog.

As part of the animacy questions, children were asked to classify each artifact as a gendered ("he" or "she") or nongendered ("it") entity. The

Table 7.1
Percentage of Children Who Provided Affirmative ("Yes") Responses to Evaluation Questions by Artifact and Age

Evaluation question	AIBO			Stuffed Dog		
	Y	O	C	Y	O	C
Animacy (n = 80)						
1.1 Is X alive?	36	41	38	40	21	30
1.2 Can X die?	15	13	14	10	15	13
1.4 Is X a real dog?	32	13	22	28	8	18
Biological properties (n = 40)						
2.1 Does X have a stomach?	58	85	72	70	85	78
2.2 Do you think X will eat this dog biscuit?	78	30	53	70	30	50
2.3 Does X grow bigger?	45	25	35	55	26	41
2.4 Does X pee and poop?	30	32	31	45	25	35
2.5 Does X breathe?	47	25	36	45	30	38
2.7 Can X have babies?	53	45	49	53	35	45
Mental States (n = 40)						
3.1 Do you think X will try to get the toy?	80	76	78	80	53	67
3.2 Can X feel happy?	79	70	74	90	60	75
3.3 Do you think X can see the doggie toy?	75	61	68	80	50	65
3.4 "Come here X." Do you think X can hear me?	61	30	45	70	25	48
Social rapport (n = 40)						
4.1 Do you like X?	74	95	85	75	95	85
4.2 Do you think X likes you?	65	95	80	74	95	84
4.3 Do you think X likes to sit in your lap?	74	80	77	89	85	89
4.4 Can X be your friend?	61	90	76	72	90	82
4.5 Can you be a friend to X?	60	94	76	74	95	84
4.6 If you were sad, would you want to spend time with X?	55	70	64	63	74	68

Table 7.1
(continued)

Evaluation question	AIBO			Stuffed Dog		
	Y	O	C	Y	O	C
Moral standing (n = 40)						
5.1 [Hit X.] Do you think it's [not] OK that I hit X?	53	85	69	55*	90*	73
5.2 Do you think X feels pain?	32	60	46	42*	85*	64
5.3 Do you think it's [not] OK to leave X alone for a week?	69	80	74	60*	85*	73
5.6 Do you think it's [not] OK to throw X in the garbage?	83	89	86	79*	95*	87
5.7 Let's say X knocks over a glass of water and spills it all over the floor. Should X be punished?	37	42	39	50*	30*	40

Note. Abbreviations for the columns are defined as: Y = younger children; O = older children; and C = combined by age. Questions 5.1, 5.3, and 5.6 above were asked in their affirmative form (e.g., "Do you think it's OK that I hit X?") and have been inverted in this table such that an affirmative response indicates an attribution of moral standing.
*The group of moral standing questions marked with asterisks was the group of evaluation questions for which significant developmental differences were found.

majority of children (55%) classified AIBO as a "he," with equal numbers classifying AIBO as a "she" (23%) or "it" (23%). For the stuffed dog, children made similar classifications: "he" (42%), "she" (28%), or "it" (30%). Gender differences were found for both AIBO and the stuffed dog. For AIBO, 37 percent of females said "she" as compared to only 8 percent of males; for the stuffed dog, 44 percent of females said "she" as compared to only 11 percent of males. Thus, compared to the female children, the male children were less likely to identify AIBO or the stuffed dog as female. This gender difference was the only one found in all of our statistical analyses of the data in this study.

As part of the mental states questions to assess children's attributions of autonomous action to the artifact, children were asked a question to describe what each artifact would do when the interviewer hid the dog toy. Children's responses were coded according to the type of action they said the artifact would perform. In turn, these coded responses were analyzed in terms of two overarching categories: "do nothing" and "do something" where the "do something" categories entailed (a) try to get the toy, (b) eat the toy, (c) play with the toy, and (d) verbally engage with the toy. Results showed children more often ascribed autonomous action ("do something") to AIBO than to the stuffed dog.

Overall, 51 percent of the children provided justifications for their evaluations (older children 63%, younger children 38%). Children's justifications were coded based on 7 overarching categories and 25 subcategories. The seven overarching categories were artifactual, biological, pretense, mental, social, moral, and child's interests. (See Kahn et al. 2006 for a description of each category; and Kahn et al. 2003 for the entire detailed justification coding system.) The key result for our purposes here is that when children provided justifications, they used virtually identical justifications for AIBO and the stuffed dog to support their positive (yes) evaluations.

Observed Behavioral Interactions

Children's behavioral interactions with the artifacts were coded with the six overarching categories and 22 subcategories as shown and defined (with examples and a still image from the video data) in table 7.2. The six overarching categories are *exploration, apprehension, affection, mistreatment, endowing animation,* and *attempt at reciprocity.* Coding reliability was established at the more nuanced level of the subcategories. In total, 2,360 behavioral interactions were coded, 1,357 with AIBO (58%) and 1,003 with the stuffed dog (43%). As mentioned earlier, any repeated or continuous behavior that occurred within each minute was coded only once within that minute segment. Thus, the 2,360 coded behaviors represent a lower bound on the children's actual behaviors.

Table 7.3 reports the total, mean, and maximum number of occurrences of each behavior for AIBO and the stuffed dog. In contrast to the results from the interview data, pervasive differences were found in children's behavioral interactions with AIBO and the stuffed dog. Statistically significant differences were found in 15 of the 22 behavioral subcategories. As shown in table 7.3, with AIBO children more often

Table 7.2
Coding Categories for Behavioral Interactions

Behavioral category	Definition and example	Still image from video

1. Exploration
  1.1 Anatomy check
  1.2 Touch limbs
  1.3 Demonstrate
  1.4 Feed
	Reference to the child's visual or tactile exploration, manipulation, inspection, pointing, and feeding of the artifact (e.g., child explains to the interviewer that AIBO is a boy while inspecting the hindquarters of AIBO).	

2. Apprehension
  2.1 Startle
  2.2 Wariness
	Reference to the child exhibiting a startle response, wariness, or other intentional movement away from the artifact (e.g., AIBO stands and child backs away quickly).	

Table 7.2
(continued)

Behavioral category	Definition and example	Still image from video
3. Affection 3.1 Nonexpl. touch 3.2 Pet 3.3 Scratch 3.4 Kiss 3.5 Embrace 3.6 Verbal	Reference to the child engaging in petting, scratching, kissing, carrying, embracing, and one-way verbal greetings to the artifact (e.g., child squeezes the stuffed dog in a big hug).	
4. Mistreatment 4.1 Rough handling 4.2 Thumping 4.3 Throwing	Reference to the child's behavior showing disregard for the artifact, including rough handling (e.g., hitting, squishing) and throwing (e.g., child swings the stuffed dog overhead and then thumps it to the floor).	

Table 7.2
(continued)

Behavioral category	Definition and example	Still image from video
5. Endow animation 5.1 Vocalize 5.2 Movement 5.3 Object play 5.4 Feed	Reference to the child enlivening the artifact in order to perform a behavior or action with it, including making sounds and moving the artifact around (e.g., child throws the bone and says "Fetch!" then child picks up the stuffed dog and begins to hop it toward the toy).	
6. Attempt reciprocity 6.1 Motion 6.2 Verbal 6.3 Offering	Reference to the child's behavior not only responding to the artifact, but expecting the artifact to respond in kind based on the child's motioning behavior, verbal directive, or offering (e.g., AIBO is searching for a ball; child observes AIBO's behavior and puts the ball in front of AIBO and says, "Come get it!").	

Table 7.3
Comparison of Frequency of Children's Observed Behavioral Interactions Directed Toward AIBO and the Stuffed Dog

Behavioral category	AIBO			Stuffed dog			Wilcoxon p-value
	Mean	Total	Max	Mean	Total	Max	
1. Exploration	2.76	221	9	1.88	150	7	.013
1.1 Anatomy check	0.06	5	1	0.09	7	2	.593
1.2 Touch limbs	1.91	153	9	1.00	80	5	<.0005*
1.3 Demonstrate	0.35	28	4	0.71	57	4	.004*
1.4 Feed	0.46	37	4	0.14	11	3	.001*
2. Apprehension	1.79	143	11	0.01	1	1	<.0005*
2.1 Startle	0.70	56	6	0.00	0	0	<.0005*
2.2 Wariness	1.44	115	11	0.01	1	1	<.0005*
3. Affection	3.67	294	14	3.87	310	17	.659
3.1 Nonexpl. touch	0.09	7	1	0.06	5	1	.527
3.2 Pet	2.10	168	8	1.72	138	8	.155
3.3 Scratch	0.15	12	1	0.20	16	3	.540
3.4 Kiss	0.05	4	3	0.04	3	1	1.000
3.5 Embrace	0.60	48	5	1.80	144	10	<.0005*
3.6 Verbal	0.81	65	6	0.14	11	2	<.0005*

Table 7.3
(continued)

Behavioral category	AIBO			Stuffed dog			Wilcoxon p-value
	Mean	Total	Max	Mean	Total	Max	
4. Mistreatment	0.49	39	7	2.30	184	15	<.0005*
4.1 Rough handling	0.49	39	7	1.90	152	11	<.0005*
4.2 Thumping	0.00	0	0	0.24	19	6	.011
4.3 Throwing	0.00	0	0	0.16	13	4	.005*
5. Endow animation	0.25	20	5	2.59	207	18	<.0005*
5.1 Vocalize	0.18	14	5	0.80	64	12	.002*
5.2 Movement	0.05	4	1	1.21	97	12	<.0005*
5.3 Object play	0.05	4	2	0.80	64	7	<.0005*
5.4 Feed	0.01	1	1	0.43	34	4	<.0005*
6. Attempt reciprocity	8.54	683	32	2.25	180	14	<.0005*
6.1 Motion	0.05	4	2	0.00	0	0	.102
6.2 Verbal	0.67	54	10	0.14	11	4	.003*
6.3 Offering	8.08	646	28	2.19	175	13	<.0005*

Note. Wilcoxon signed rank test is a nonparametric test for comparing two related distributions. It does not compare means, but the means are reported in this table for their descriptive value.
*Indicates behavior categories for which there were statistically significant differences between AIBO and the stuffed dog after adjusting for multiple comparisons using Holm's Sequential Bonferroni method.

engaged in apprehensive behavior and attempts at reciprocity. In contrast, with the stuffed dog, children more often engaged in mistreatment and endowing animation.

Children's behavioral interactions were also analyzed in terms of their co-occurrence with the artifact-initiated stimuli (which was always AIBO since the stuffed dog never initiated its own action), interviewer-initiated stimuli, and time-linked stimuli. A total of 1,689 behavioral co-occurrence stimulus–behavior dyads occurred. Results showed that virtually all of the children's apprehensive behaviors (134 observations, 99%) were observed in conjunction with AIBO after AIBO initiated a behavior, particularly when AIBO moved in place (38 observations, 28%) and approached the child (59 observations, 44%). In contrast, virtually all of the children's behaviors that involved mistreatment (95 observations, 79%) and endowing of animation (126 observations, 89%) were observed in conjunction with the stuffed dog, predominantly after the interviewer had engaged in verbal behavior (mistreatment, 82 observations, 68%; endowing animation, 104 observations, 74%). Children most frequently attempted reciprocal interactions after AIBO-initiated behaviors (352 occurrences, 59%), secondarily after interviewer-initiated behaviors with AIBO (163 occurrences, 27%), and to a lesser degree after interviewer-initiated behaviors with the stuffed dog (86 occurrences, 14%). In addition, a total of 671 behaviors were coded where no preceding stimulus within five seconds of the behavior could be discerned. In this context children engaged in more affection (107 occurrences, 66%), mistreatment (88 occurrences, 89%), and endowing of animation (81 occurrences, 94%) with the stuffed dog than with AIBO. In contrast, children engaged in more attempts at reciprocity (153 occurrences, 63%) with AIBO than with the stuffed dog.

In terms of children's responses to witnessing the interviewer hit (sharply tap) the artifact on its head, two types of data were coded for the first five seconds immediately following the hit: whom or what the child looked at and how long the child looked. Results showed immediately following the hit that most children looked at the artifact (AIBO 95%, stuffed dog, 79%). Among the children who looked at the artifact first for both AIBO and the stuffed dog, on average children looked at AIBO 2.3 times longer than they looked at the stuffed dog, a statistically significant difference.

Card Sort Task

Children's responses to the card sort task were tested for each of the six comparisons—AIBO compared with pairs chosen from desktop computer, robot, stuffed dog, and real dog. For each pairwise comparison, a statistical test was used to look for evidence of a clear preference among the children as to which of the two objects was more similar to AIBO. Results showed that a significant majority of children viewed AIBO as less like a desktop computer and more like a robot (74%), a stuffed dog (75%), or a real dog (67%). Overall, children were approximately evenly split on the robot versus the stuffed dog, the robot versus the real dog, and the real dog versus the stuffed dog. When age differences were tested, one was found: Older children (62%) were more likely than younger children (33%) to view AIBO as more like a stuffed dog than a real dog.

Conclusion

This study examined preschool children's reasoning about and behavioral interactions with what, at the time, was the most computationally sophisticated robotic pet on the retail market, Sony's robot dog AIBO. Results showed that one quarter of the children accorded animacy to AIBO, about one half of the children accorded biological properties to AIBO, and about two-thirds of the children accorded mental states, social rapport, and moral standing to AIBO. However, children provided virtually the same proportion of such evaluations to the stuffed dog. Similarly, in supporting their positive evaluations, children provided virtually identical justifications about AIBO and the stuffed dog. Thus one interpretation of these results is that these children engaged in imaginary play with AIBO in the same way and to the same degree that they engaged in imaginary play with the stuffed dog.

Yet this interpretation is called into question by the behavioral results. Most notably, based on an analysis of 2,360 coded behavioral interactions, children engaged more often in apprehensive behaviors and attempts at reciprocity with AIBO. In contrast, children more often mistreated the stuffed dog and endowed it with animation. Thus, taking both the reasoning and behavioral results together, it is possible that children made identical judgments about AIBO and the stuffed dog, but actually believed the former but not the latter.

I think that happened often. For example, about half the children said that AIBO and the stuffed dog could hear; if they really believed it then one would have expected children to use verbal directives to AIBO and the stuffed dog in about equal proportion. However, the results showed that children used more verbal directives to AIBO (54 occurrences) than to the stuffed dog (11 occurrences). Similarly, about half the children said that AIBO and the stuffed dog could feel pain; if they really believed it then one would have expected the children to seldom mistreat either arti-fact or (if for some reason this group of children did not care about hurting a sentient creature) to mistreat both artifacts proportionately. However, the results showed that children often mistreated the stuffed dog (184 occurrences) but seldom mistreated AIBO (39 occurrences). Also, children often flinched away from AIBO immediately after AIBO initiated an action (e.g., standing, walking, or approaching the child). This apprehensive behavior is evidence that the children believed that AIBO could be a threat. Indeed, virtually all of the children's apprehensive behaviors occurred when AIBO initiated action and almost never when the interviewer initiated an interaction with AIBO. This pattern might well mimic that of children in the presence of an unfamiliar live dog—a little apprehensive when the dog owner is not controlling the dog, and not apprehensive when the dog owner is in control. Finally, results showed that children often animated the stuffed dog (207 occurrences) but not AIBO (20 occurrences). It is as if the children expected that AIBO had the ability to direct its own behavior, and did not need their assistance.

The results from the card sort task further support the proposition that AIBO was not conceptualized as strictly an inanimate artifact. Based on all pairwise comparisons, results showed that children viewed AIBO as less like a desktop computer and more like a robot, a stuffed dog, or a real dog. In other words, the children did not categorize AIBO as either more like an animate or inanimate entity, even though preschool-aged children in general are quite good at doing so with prototypic objects (Gelman, Spelke, and Meck 1983; Gelman and Markman 1986).

In summary, we had set out in this study to show decisively that pre-school children were not pretending when they engaged socially and morally with AIBO. That was the reason methodologically that we had introduced the stuffed dog as the comparison artifact. Did we succeed? Not decisively, because children's judgments did not show any difference between the two artifacts. Instead, we had to draw on our rich and detailed behavioral analyses and card sort data to suggest that children were not pretending in their judgments about AIBO and were pretending in their judgments about the stuffed dog. One goal, then, for future research is to try to gain decisive evidence that children are not pretending when they engage socially and morally with robot pets. One approach would be to devise methods that readily move children "into" and "out of" pretend spaces, and then test experimentally whether it is possible to move children into and out of pretense with a stuffed dog but not with AIBO.

Data aside, the larger conceptual issues of this study are also central to understanding the human relation with technological nature. When people engage with technological nature, do they think it is real? Consider again the plasma display studies from chapters 4 and 5. Was the real-time video feed of the local nature scene a real view? My answer is that it was not real in the sense of being identical to looking at the nature scene without any technological mediation. But it was real in that it was a real real-time video feed of the nature scene. The reader might respond that I have not yet answered much. But that is partly my point. Would the view be real if seen through eye glasses? Would the view be real if seen by a person who sees with 20–50 vision or with advanced glaucoma? Is the voice you hear through a phone landline or through a cell phone a real voice? What if that same voice was first processed through a voice synthesizer to alter its pitch? Is digitized music real music? Is it less real than live music? Is live music less real if it is first amplified to project its sound into a large auditorium? For some of these questions, people might not be sure how to answer. The reason is that the parameters are not well specified for what makes something real.

The question of what is real will probably become even more confusing in the decades ahead with personified robots. The reason is that this form of technological nature will become increasingly responsive to us in social terms. Thus, we might well start living our lives largely committed in action, if not in belief, about their social nature. At the same time, we might be committed to a belief that the entities are technological. We saw evidence of exactly this constellation of commitments and beliefs in the previous chapter wherein the majority of the members of the discussion forums wrote compellingly of how they viewed AIBO simultaneously as a technological and social other.

8

Robotic Dogs and Their Biological Counterparts

We learned in chapter 6 that members in the AIBO online discussion forums often seemed to relate to their robotic dogs as if they were biologically live dogs. Recall, for example, the member who wrote: "The other day I proved to myself that I do indeed treat him as if he were alive, because I was getting changed to go out, and [AIBO] was in the room, but before I got changed I stuck him in a corner so he didn't see me! Now I'm not some socially introvert guy-in-a-shell, but it just felt funny having him there!" Or recall the member who wrote: "Oh yeah I love Spaz [the name for this member's AIBO], I tell him that all the time. . . . I care about him as a pal, not as a cool piece of technology. I do view him as a companion, among other things he always makes me feel better when things aren't so great." This sort of data is suggestive of a future—not so unlike the story I summarized in chapter 6 of *Do Androids Dream of Electric Sheep*—where robot animals have a powerful social presence in human lives and could become reasonable replacements for biologically live animals.

But what would we find if we compared people's reasoning about and behavioral interactions with AIBO directly with their reasoning about and interactions with a biologically live dog? With such a direct comparison, how would AIBO match up with its biologically live counterpart? That question structured the current study my colleagues and I conducted. The lead researcher was Gail Melson at Purdue University (Melson Kahn, Beck, Friedman, Roberts, Garrett, et al. 2009).

Seventy-two children participated in this study, equally divided between three age groups (7–9 years, 10–12 years, and 13–15 years). There were equal numbers of girls and boys. Most children came from families that had pets: 87 percent had at least one pet at the time of the study, and all but two families had a history of pet-keeping. Dogs were the most common pet, with 89 percent of pet owners having at least one dog. None of the participating children owned or had played with AIBO previously, although 37 percent owned some type of simple "robotic pet," such as Technodog or Furby. All the children were regularly using a computer either at home or in school and usually in both.

We used the same model AIBO (the 210 version) as we used in the study of preschool children reported in chapter 7. The biologically live dog, referred to as Canis, was one of two Australian Shepherds, a mother and her daughter. It was difficult to tell the two dogs apart from one another in looks and behavior. These dogs were Delta Society certified to serve in animal-assisted activities (AAA) involving children.

Three major research questions structured our study: First, do children behave toward a robotic dog in ways that are similar to how they behave with a biologically live dog or with a complex artifact? Based on the study reported in chapter 6, we expected to uncover evidence for both. Second, in terms of their conceptualizations, to what extent do children affirm or deny the biology, psychology, social companionship, and moral standing of a robotic dog as compared to a biologically live dog? There is considerable evidence (for review, see Melson 2001) that children as well as adults believe that dogs and other pets, while recognized as clearly nonhuman, nevertheless possess unique psychological attributes, have feelings, desires, and cognition, are capable of social companionship, and should be accorded moral standing (Beck and Katcher 1996). Even preschoolers experience living animals as subjective others, with minds and emotions (Myers 1998). We expected that most children, even the youngest in our study, would perceive marked differences between AIBO and the biologically live dog, according the former but not the latter artifactual properties. However, it was an open question the extent to which children would conceive of AIBO as biological, psychological, social, and moral. Third, are there age group differences in children's behaviors and understandings? Based on limited prior research (Bryant 1985; Furman 1989), we expected that children in the youngest age group (7- to 9-year-olds) would treat AIBO as more of a social companion than would older children. We did not expect age group differences in children's conceptions about the biology or psychology of AIBO, given the establishment of causal-explanatory reasoning systems for naive biology and physics by age 7 (Wellman, Hickling, and Schult 2000).

Each child participated in two individual sessions, one with AIBO and one with Canis, each approximately 45 minutes long. Order of sessions was counterbalanced. Each session began with a five-minute unstructured play period with the target dog and a pink plastic ball. A 38-question interview followed, with the child free to continue to play with the target dog. During the interview, the child responded to a series of questions tapping the child's conceptions of the target dog in the four domains established in the AIBO studies reported in chapters 6 and 7: (a) biological entity (6 questions); (b) mental states (10 questions); (c) social companion (15 questions); and (d) moral standing (6 questions). Each question asked the child to affirm or deny some characteristic of the target dog. For *biological entity*, each question asked about a biological process or feature, for example: "Does X eat?" For *mental states*, each question asked about a psychological state ("Can X feel embarrassed?"), a sensory capacity ("Can X hear you?"), or a cognitive understanding ("Can X understand you?"). Two questions in this domain asked about the child's readiness to communicate with the target dog: "Would you talk to X?" and "Would you tell secrets to X?" For *social companion*, questions were of two types: X as a social companion to the child, for example, "Can X be your friend?" (8 questions), and the child as a social companion to X, for example, "Can you be a friend to X?" (7 questions). For *moral standing*, six questions asked the child if it was OK or not OK to engage in a series of actions (from less to more severe) that would harm X. Note that the affirmative answer, "OK," indicates *lack of* moral standing. Order of domain was counterbalanced across participants; but within domain, the order of questions was fixed. Pilot testing suggested an optimal within-domain question order that avoided repetition and encouraged the child to elaborate his or her thinking (in follow-up questions). For example, in the moral standing domain, questions progressed from less to more severe harmful actions against the target dog. See table 8.1 for complete set of questions.

Following the second session, after removing the target dog, we conducted a card sort task. In this task, the interviewer presented a laminated 4 × 6 color photograph of AIBO. With this card as the anchor, all pair-wise comparisons of the following color photographs were shown:

Table 8.1
Interview Protocol (X = Robotic Dog, Biologically Live Dog)

Domain	Question
Biological entity	Does X have a stomach?
	Does X eat?
	Does X go to the bathroom? (pee or poop)
	Can X get sick?
	Can X die?
	Can X have babies?
Mental states	Can X feel happy?
	Can X feel embarrassed?
	Can X see the ball?
	Can X see you?
	Can X hear you?
	Does X understand you?
	If you saw X again, would X recognize you?
	Does X know how you are feeling?
	Would you talk to X?
	Would you confide in X?
Social companion	Do you like X?
	Does X like you?
	Can X like anyone X wants?
	Can X be your friend?
	Can you be a friend to X?
	If you were sad, would you feel better with X?
	If X were sad, would X feel better with you?
	Can you play with X?
	Can X play with you?
	If a friend came over and you were playing with your friend, would X feel left out?
	If a friend were playing with X, would you feel left out?
	If you were going to sleep, would you want to cuddle with X?
	If X were going to sleep, would X want to cuddle with you?
	If you were home alone, would you feel better with X?
	If X were home alone, would X feel better with you?

Table 8.1
(continued)

Domain	Question
Moral standing	If X were whimpering (making noise) would it be OK or not OK to ignore X? If X's leg breaks, is it OK or not OK to not fix it right away? If you decided you did not like X anymore, is it OK or not OK to give X away? If you decided you did not like X anymore, is it OK or not OK to throw X in the garbage? If you decided you did not like X anymore, is it OK or not OK to destroy X? Is it OK or not OK to hit X?

humanoid robot, live dog, stuffed dog, and desktop computer. Thus, the following pair-wise comparisons were shown, in random order: robot/biologically live dog; robot/stuffed dog; robot/desktop computer; desktop computer/biologically live dog; desktop computer/stuffed dog; and stuffed dog/biologically live dog. For each A, B pair, the child was asked, "Is AIBO more like A or more like B?"

Both sessions plus the card sort task were videotaped from behind a one-way mirror. The child wore a wireless microphone, and a supplemental conference microphone in the room recorded all audio.

Children's behaviors toward the target dog were coded from the videotaped (5-minute) play portion of each session. Children's interview and card sort responses were coded from transcripts of each session derived from both the videotapes and audiotapes. Intercoder reliability was established.

Observed Behavioral Interactions

Based on an inductively developed ethogram (Montagner et al. 1988) of children's behaviors toward the target dog, we coded for three types of social behaviors: social touch (gentle pats, petting, scratching, kissing, hugging), verbal engagement (greetings, general verbalizations), and attempts at reciprocity, both verbal (asking questions, commanding) and nonverbal (motioning, offering the ball, present one's hand). In addition, we coded the child's exploration of the target dog (for example, poking in an effort to determine how the target moved) and proximity to the target dog, defined as being within the child's arm reach. Finally, because children were interacting with a novel object (robotic dog) or unfamiliar dog, we coded any apprehensive behaviors, such as startle responses, withdrawals, or fear expressions. The onset and offset of each behavior of interest were recorded; an instance was defined as offset equal to or greater than two seconds.

Table 8.2 presents the total number of instances of each behavior for the entire sample, the median and range of the number of instances per child, and the number and percent of children exhibiting at least one instance of each behavior. Results showed that children explored the robotic dog as an artifact more frequently than they did the biologically live dog. In contrast, children used social touch, particularly petting and scratching, more with the biologically live dog than with the robotic dog. Median frequency of social touch behaviors was more than six times as high with the biologically live versus the robotic dog. Although 63 percent of the children spoke to the robot at least once, the frequency of social speech was higher with the biologically live dog (median = 3 occurrences per child) than with the robotic dog (median = 1). Virtually all children (96%) attempted to engage both dogs in interaction using social movements, especially offering the ball, but children offered the ball significantly more frequently to the robotic dog (median = 7 occur- rences per child) than to the live dog (median = 2). Conversely, attempts at verbal engagement were more likely with the biologically live dog; 69 percent of children asked the biologically live dog a question at least once, whereas 44 percent asked the robotic dog a question.

Children's Reasoning

During the interview we asked two types of questions. Affirmation/nega- tion questions asked the child to respond "yes" or "no" ("OK" or "not OK" for moral standing questions). Answers were coded as an affirma- tion ("yes," "OK") or a negation ("no," "not OK"). The second type of question was an open-ended follow-up to each affirmation/negation question and prompted the child for justification of the preceding

Table 8.2
Frequency of Occurrence of Children's Behavioral Interactions Toward Robotic Dog and Live Dog

Behavior	Robotic dog (n = 72)						Biologically live dog (n = 72)						Sign test p-value
	Total	Median	Range Min	Range Max	At least once n	At least once %	Total	Median	Range Min	Range Max	At least once n	At least once %	
Exploration as artifact	254	2	0	18	43	60	37	0	0	5	17	24	<.0005*
Proximity	47	0	0	5	26	36	73	0	0	9	32	44	.617
Social touch	276	2	0	22	45	63	1074	13	2	39	72	100	<.0005*
General	48	0	0	6	15	21	54	0	0	12	27	38	.216
Petting	168	0	0	17	33	46	670	8	0	24	71	99	<.0005*
Scratching	29	0	0	7	12	17	317	4	0	21	53	73	<.0005*
Kissing	1	0	0	1	1	1	5	0	0	2	4	6	.375
Hug	1	0	0	1	1	1	8	0	0	5	3	4	.625
Mistreatment	6	0	0	3	3	4	8	0	0	3	4	6	1.000
Verbal engagement	397	2	0	33	59	82	534	7	0	26	62	86	.200
Salutation	77	0	0	7	33	46	85	1	0	11	44	61	.775
General	263	1	0	31	45	63	379	3	0	24	56	78	.001*

Table 8.2
(continued)

Behavior	Robotic dog (n = 72)						Biologically live dog (n = 72)						Sign test p-value
			Range		At least once				Range		At least once		
	Total	Median	Min	Max	n	%	Total	Median	Min	Max	n	%	
Attempt at reciprocity	1267	14	0	68	69	96	924	9	0	65	69	96	.002*
Motioning	189	1	0	28	43	60	149	1	0	13	39	54	.341
Directive	301	2.5	0	22	48	67	248	1	0	20	44	61	.127
Questioning	164	0	0	19	32	44	244	2	0	23	50	69	.010
Ball offering	558	7	0	31	64	89	240	2	0	27	58	80	<.0005*
Hand presentation	55	0	0	5	29	40	43	0	0	4	23	32	.499
Apprehension	26	0	0	6	13	18	8	0	0	5	4	6	.006

Note. When totals for higher-level categories are higher than the sum of totals for subcategories, behaviors were coded at the higher level or other subcategories have been omitted.
*Indicates statistically significant differences after adjusting for multiple comparisons using Holm's Sequential Bonferroni method (family sig. level .05).

response. Thus, there were 6 justification responses for biological entity, 10 for mental states, 15 for social companion, and 6 for moral standing. Drawing on Kahn et al. (2003), a coding manual was developed that assigned each thought unit (a meaningful phrase or sentence expressing a distinct form of reasoning) to one or more of the following forms of justification in terms of category membership, attributes, or processes: artifact, biological entity, "qualified" biological entity (for example, "he [AIBO] doesn't *really* have eyes"), psychological entity, social companion, moral entity, and analogical reasoning as like a human or animal.

Results showed that children consistently affirmed the biology, psychology, social companionship and moral standing of the biologically live dog more than the robot. Specifically, for the biologically live dog, 99 percent of the children affirmed biological attributes, 83 percent mental states, 91 percent social companionship, and 86 percent moral standing. For the robotic dog, 22 percent of the children affirmed biological attributes, 56 percent mental states, 70 percent social companionship, and 76 percent moral standing.

There were no significant age group or sex differences for responses about the biologically live dog. For the robotic dog, there were no sex differences, but there were significant age group differences in social companionship and moral standing. For social companionship questions, 7- to 9-year-olds affirmed 82 percent of the questions, as compared with 71 percent of the 10- to 12-year-olds and 55 percent of the 13- to 15-year-olds. Similarly, for the moral standing questions, the mean percentage of affirmations for the youngest group was 86 percent, 76 percent for the 10- to 12-year-olds, and 64 percent for the oldest group. Though it was not quite statistically significant, a similar pattern occurred descriptively for biological properties of the robot: 26 percent for 7- to 9-year-olds, 26 percent for 10- to 12-year-olds, and 14 percent for 13- to 15-year-olds.

All but one child justified their judgments, at least once, by explaining that the robot was an artifact or possessed artifactual properties, and they tended to use this justification category quite frequently with the robot (median = 10.5 occurrences per child), but very rarely with the live dog (median = 0 occurrences per child, with only 15% ever using the category). All or virtually all children mentioned at least once that the biologically live dog was a biological entity (or possessed biological properties), had mental and emotional properties, was a social companion, and had moral standing. These results were expected; but the overwhelming majority of children also, at least once, justified their prior judgments in terms of the robotic dog's biology (99%), mental states (94%), social companionship (89%), and moral standing (92%). However, while most children did use each of these justification categories at least once with the robotic dog, they employed reasoning based on affirming biology (live dog, median = 11; robotic dog, median = 4), mental states (live dog, median = 11; robotic dog, median = 6), and moral standing (live dog, median = 6; robotic dog, median = 4) more frequently across the interview with the biologically live dog than with the robotic dog. Children also drew more frequently on analogies to other living animals and to humans with the biologically live dog (median = 5) than with the robotic dog (median = 2). However, there were no significant differences between dog types with respect to the frequency of use of justifications based on social companionship.

Card Sort Task

Results from the card sort task showed that children viewed AIBO as more like a robot than a desktop computer (87%), a stuffed dog (90%), or a biologically live dog (74%). Children also saw AIBO as more like a desktop computer than a stuffed dog (74%). Children were more closely divided on whether AIBO was more like a desktop computer than a biologically live dog (46%) or more like a biologically live dog than a stuffed dog (58%). Boys (71%) were more likely than girls (46%) to say that AIBO was more like a biologically live dog than a stuffed dog.

Conclusion

Based on some of our measures, children clearly differentiated AIBO from the biologically live dog. Notably, in the unstructured play period, the majority of children explored AIBO as an artifact, poking and touching AIBO as they would an unfamiliar toy. Only 24 percent of the children ever touched the biologically live dog in this way and they did so infrequently. Compared with the biologically live dog, children much less frequently engaged toward AIBO in gentle, affectionate social touch, particularly petting or scratching. Across the many interview questions,

children consistently affirmed the biology, psychology, companionship, and moral standing of the biologically live dog more than of AIBO. In justifying these affirmations, children drew on reasoning that AIBO belonged predominantly to the category of "artifact," possessing artifactual properties and processes, whereas the biologically live dog belonged to the category of biological entity and mental agent. In addition, children drew on analogies to other living animals and to humans more frequently when explaining responses about the biologically live dog compared to AIBO. Finally, based on the card sort task, children viewed AIBO more like a robot than a biologically live dog or even a stuffed dog.

That said, it was also the case that the majority of children treated AIBO in ways that were doglike. In other words, although the differences in children's evaluations between AIBO and the biologically live dog across categories were statistically significant, a surprising majority of children affirmed that AIBO had mental states (56%), sociality (70%), and moral standing (76%). In addition, while children spent more time within arm's reach of the biologically live dog than AIBO, most children (80%) spent the majority of their time within arm's reach of both. As one child said, when asked how she would play with AIBO, "I would like play with him and his ball and just give him lots of attention and let him know he's a good dog." When one considers AIBO's hard metal surface, it is also noteworthy that 46 percent of children gently petted the robot at least once.

Taken together, the results show that children believed that AIBO was a technology to which they accorded many of the core attributes of a minded animal. What is going on? How can children think that AIBO is not a live dog and yet that it is?

I would like to offer a speculative answer. I believe a new technological genre is emerging that challenges, and will increasingly challenge, our existing cognitive categories, between, for example, "alive or not alive," "animate or inanimate," "having agency or not," or "being a social other or not." This interpretation is congruent with the data from the previous two chapters of AIBO in the lives of adults (chapter 6) and preschool children (chapter 7). Even more speculatively, I believe that in the future annals of human history this technological genre will emerge as a new ontological category in human minds. By an ontological category I mean a fundamental category of human existence that each person constructs through interaction with a physical and social world, by means of a developmental process that involves the equilibration of cognitive structures (Kahn 1999; Piaget 1983; Turiel 1983). Such categories include time, space, dead, alive, number, logic, and morality.

If this speculation is correct, then it is likely that the English language is not yet well equipped to characterize or talk about it. As an analogy, we do not normally present people with an orange object and ask, "Is this object red or yellow?" It is something of both, and we call it orange. Similarly, it may not be the best approach to keep asking if animal robots are, for example, "alive" or "not alive" if, from the person's experience of the subject–object interaction, the animal robot is alive in some respects and not alive in other respects, and is experienced not simply as a combination of such qualities (in the way one can inspect a tossed salad and analytically distinguish, for example, between the green leaf lettuce and the red leaf lettuce) but as a novel entity. If that is the case, then researchers (myself included) will continue to encounter difficulties as long as we continue to ask questions based on the only language and categories we know to investigate an entity where the answers are "both" and "neither."

Robotic Dogs Might Aid in the Social Development of Children with Autism

Throughout much of this book I have been showing how technological nature is better than no nature but not as good as actual nature. In this chapter, I focus on the first part of this proposition. I discuss a study my colleagues and I conducted to investigate whether a robotic dog (AIBO) could aid in the social development of children with autism. If the answer is yes, it would highlight an important application for technological nature.

Before turning to what my colleagues and I did and what we found, I would like to say a little about what autism is and what has been done to date with children with autism interacting with robots.

Briefly, autism is a pervasive developmental disorder characterized by three main symptoms (Dawson and Toth 2006; Koegel, Koegel, and McNerney 2001; Rogers 2006). The first—and perhaps most significant—is impaired social interaction. Compared to their typically developing peers, children with autism focus their attention more on objects, such as toys, and interact much less with other people. This lack of interaction leaves children with autism with poor social skills, which in turn sets into motion a cycle where their awkward attempts at social interaction meet with negative feedback from other children and adults alike, which leads to further social withdrawal. The second symptom is impaired communication. Approximately 30 percent of individuals with autism lack spoken language. Children with autism who can speak are likely to demonstrate atypical mannerisms, such as immediate or delayed echolalia, which is the tendency to mimic others' speech rather than forming their own sentences. Other atypical verbal patterns include unusual intonation, syntax, and word choice. The third symptom is the presence of repetitive behaviors and restricted interests. Repetitive behaviors may take the form of repeated motor movements, such as hand-flapping, toe-walking, spinning in circles, or even self-injurious behaviors, as well as persistently using objects in a nonfunctional, repetitive manner (e.g., spinning the wheels on a toy car for an hour).

In the field of human–robot interaction (HRI), there have been some pioneering studies that suggest that robots might be able to ameliorate one or more of these symptoms (Dautenhahn and Werry 2004; Dautenhahn et al. 2003; François, Polani, and Dautenhahn 2007; Kozima et al. 2004; Kozima, Nakagawa, and Yasuda 2005; Kozima and Yano 2001; Liu et al. 2007; Robins and Dautenhahn 2006; Robins, Dautenhahn, Boekhorst, and Billard 2004ab, 2005; Robins, Dickerson, Stribling, and Dautenhahn 2004; Scassellati 2005, 2007; Shibata et al. 2005). A large body of this research was initiated by Dautenhahn, Robins, and their colleagues. For example, they conducted a number of studies of children with autism interacting with a mobile nonhumanoid robot. They found that children were not afraid of the robot, and were more attracted to the mobile robot than a nonrobotic toy. In addition, some of the children with autism used the robot as a medium for social interaction with another child, and there were instances where children used the robot as the means to make contact with an adult. In other studies, Dautenhahn, Robins, and their colleagues used a humanoid robotic doll "Robota." In one longitudinal study, for example, they investigated the interactions of four children with autism interacting with Robota over a period of several months (Robins et al. 2004a). Two of the four children showed an increase across sessions in their overall level of interaction with the robot, as measured by the quantity of time spent looking at the robot, being in close proximity to the robot, touching the robot, and imitating the robot. Qualitative results were also suggestive. For example, one boy (with no verbal language) responded to Robota's movements with a dance of his own. Six months after the study was completed, the boy again encountered Robota, and again responded with a dance. The researchers speculated that this dance was the child's attempt to communicate with the robot. Along related lines, Kozima and colleagues have provided suggestive evidence with two different types of robots—an upper-torso humanoid robot (Kozima et al. 2004) and a small creature-like robot, "Keepon" (Kozima, Nakagawa, and Yasuda 2005)—that robots have the potential to engage children socially and provide a pivotal role for enhanced communication with adults.

My colleagues and I sought to extend this emerging body of research in three ways. The lead researcher for this study was Cady Stanton (Stanton et al. 2008). First, although researchers have been clearly aware that the robot form matters greatly in human–robot interaction (MacDorman 2005; MacDorman and Ishiguro 2006), except for one study (François, Polani, and Dautenhahn 2007), which focused more on technical issues of the robot itself, we know of no research that has investigated children with autism interacting with a robotic animal. Yet in the autism literature, animals have been shown to be effective in increasing social interaction and communication in this population. For example, in one study (Sams, Fortney, and Willenbring 2006), 22 children with autism, over the course of 15 weeks, participated in both a weekly traditional occupational therapy session and one that incorporated animals. In the sessions incorporating animals, children interacted more socially and used more language. In another study (Martin and Farnum 2002), children with autism in the presence of a dog were more likely to demonstrate a happy, playful mood, and interacted socially with the dog, both looking at it and talking to it. Moreover, in the presence of a dog the children's communications with the therapist were more likely to be on topic and interactive. Thus, for our robot in the current study we used the same animal robot used in the studies reported in the previous chapters, AIBO.

The second way we sought to extend the emerging body of HRI research of children with autism was by comparing children's social interactions with AIBO to a mechanical but nonrobotic dog, which we named Kasha (see figure 9.1). The main difference between the two artifacts was that unlike AIBO, Kasha had no ability to detect or respond to its physical or social environment. Thus this current study directly investigated whether the seemingly autonomous, self-directing, and self-organizing features of AIBO could account for improved social interaction in children with autism.

Third, we sought to deepen and coalesce a set of measures for assessing child–robot social interaction of children with autism. In our study, children interacted with both AIBO and Kasha during different times within the same experimental session. During each session, an

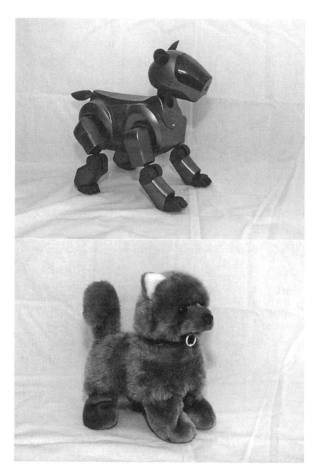

Figure 9.1
The two artifacts: AIBO (top) and Kasha (bottom).

experimenter was present and sought to engage the child in some social conversation with each of the artifacts. We then measured four aspects of social interaction that are central to the autism literature and HRI literature discussed above. The first was the amount of time children spent interacting with each artifact (AIBO and Kasha). The second was the amount of speech each child produced in talking to each artifact. The third was the number of behavioral social interactions typical of children without autism that each child engaged in with each artifact. And the fourth was the number of behavioral social interactions typical of children with autism that each child engaged in with each artifact.

Although this study was exploratory, we expected to find evidence that AIBO, as compared to Kasha, engaged children more as a social other, and facilitated greater social interaction with other people.

Eleven children between the ages of 5 and 8 participated in this study. All participants had a formal diagnosis of autism, as well as some verbal ability. In addition, participants had no significant vision, hearing, or motor impairments; no history of head injury; and no history of neurological disease. The study was open to both males and females, but only one girl enrolled in the study. This gender imbalance likely occurred, in part, because autism is three to four times more common in boys than girls (Dawson and Toth 2006).

Each child participated in an individual interactive session with both artifacts in a large room. The presentation order of the two artifacts was counterbalanced. The protocol was designed to be completed within 30 minutes, so as not to overtax the attention span of the children. Sessions were videotaped, and the camera person was present in the same room with the child.

Because children with autism often feel anxious in unfamiliar environments and with unfamiliar people, we paid special attention to making the child feel comfortable. For example, each session began with a brief introductory period designed to allow the child to become familiar with and at ease in the room, with neutral toys on the floor so the child could play with them if desired. Also, the child's parent was present in the room at all times. We asked the parent not to initiate interactions with his or her child, but said it was fine to be responsive if the child initiated such interactions. Once the parent confirmed that the child appeared comfortable in the room, the experimenter introduced the first artifact. At this junction, the experiment formally started.

The experimenter brought out of a cupboard either AIBO or Kasha (depending on the counterbalanced condition), and turned on the artifact within the child's view. The experimenter told the child it was okay to pet the artifact, and asked him if he would like to do so. The experimenter then let the child engage in self-directed exploratory play for a few minutes. Next the experimenter asked a series of preestablished questions, while the child continued to play, and invited the child to engage in certain preestablished behavioral interactions with the artifact, such as holding it and rolling the ball to it. The purpose of the questions and invited behavioral interactions was to provide a consistent structure across sessions, while encouraging social engagement with both the artifact and the experimenter.

Coding Categories and Procedures

Drawing from previous work on typically developing preschool children's interactions with AIBO reported in chapter 7, we developed a behavioral coding manual from a portion of the data and then applied it to the entire data set. The behavioral categories (and subcategories) are listed below. Key coding procedures are also described.

Through our procedures, we sought to deal with a difficult problem in coding certain types of behavioral interactions: knowing when one behavior ends and another begins. For example, imagine a child petting AIBO by running his hand back and forth along AIBO's body. Should each coupling of a back-and-forth movement be counted as "one pet?" Or should each unidirectional movement be counted as a pet? Now imagine that the child stops petting for an instant (say, half a second), and then continues petting in the same direction he had been moving. Should the movement following the slight pause be counted as the continuation of the initial petting behavior? If so, what if the child stops for one second? Five seconds? Where does a pause indicate a break in one unit of behavior and the start of a new unit of identical behavior? This example illustrates just one of dozens of such difficulties that arose. Thus, to establish a reliable means of coding a distinct behavioral unit, we coded a behavior only once within a specified unit of time: either 5 seconds or 30 seconds, depending on the specific behavior. We determined these units by reviewing video footage with the intent to find the best compromise between demarcating behavior cohesively (e.g., not chopping up a clearly single behavior into different timed sections) and increasing the frequency count of the behaviors (and thus increasing the statistical power of our analyses).

Interaction with Artifact   This category captured the time the child was engaged with the artifact, including such activities as the following: touching, talking to, throwing or kicking or offering the ball to, and gesturing to. Eye gaze was not enough to initiate an interaction; however, eye gaze during an ongoing interaction was sufficient to maintain that interaction. During a period of interaction, noninteraction could occur for up to five seconds, and still be part of the initial interaction period. If there was a break in the interaction for 5+ seconds, noninteraction started at the time of the break.

Spoken Communication to Artifact   This category captured the number of meaningful words said by the child to the artifact. Hyphenated words (e.g., "uh-huh," "choo-choo," "chow-chow") counted as one word. If the child made a number of utterances in a row, with only a subset recognizable, then only the recognizable words were coded.

Behavioral Interactions with Artifact Typical of Children without Autism   This category comprised five subcategories: (1) Verbal engagement, including salutations, valedictions, and general conversations. (2) Affection, including petting, scratching, tender touching, kissing, and hugging. (3) Animating artifact, including moving the artifact's body or part of its body (e.g., helping the artifact walk or eat its biscuit), and speaking words or sounds for the artifact (e.g., "woof-woof"). (4) Reciprocal interaction, including motioning with arms, hands, or fingers to give direction to the artifact, verbal cues (e.g., directives or questions), and offerings (e.g., showing the artifact the dog biscuit or throwing the artifact the ball)—all with some expectation of a response from the artifact. (5) Authentic interaction, either dyadic with the artifact or triadic between the child, artifact, and experimenter (e.g., the speed, tone, and volume of the child's voice is exceptionally well modulated for the circumstances, or the child's body is in a state of repose oriented toward the artifact as a social partner). Each behavior was coded only once within a 30-second period. Thus, for example, if a child petted AIBO more or less continuously for 100 seconds, that behavior would be coded as 4 occurrences of affection.

Behavioral Interactions with the Artifact Typical of Children with Autism   The Gilliam Autism Rating Scale (Gilliam 1995) lists 42 behaviors that are indicative of autism. We reviewed this inventory prior to coding the data, and pulled out all behaviors (16 in total) that we thought could be found in our data set and reliably coded. These behaviors were the following: (1) rocks back and forth, (2) flicks fingers or hands, (3) high-pitched noise, (4) unintelligible sounds, (5) repeats words, (6) lines up objects, (7) prances/walks on tiptoe, (8) whirls/turns in circles, (9) inappropriate pronouns, (10) uses third person for self,

(11) withdraws/standoffish, (12) unreasonable fear, (13) licks objects, (14) smells objects, (15) lunging/darting, and (16) self-injurious. To help the reader understand what we excluded, here are some examples (with our reasons in parentheses): "Avoids asking for things he or she wants" (would require a long-term relationship with the child to determine); "resists physical contact from others" (not appropriate for our data, as we never sought to establish physical contact with the child); "looks through people" (not possible to code reliably with our video data); and "responds inappropriately to simple commands" (we did not consistently offer any commands to the child; also it was a novel situation, so it is not clear what "appropriate" behavior would be). Each behavior was coded only once within a 30-second period. Note that I speak of the coded behaviors as "interaction with the artifact" because the entire experimental session was constituted to involve interaction between the child and the artifact, although some of the coded behaviors (e.g., "flicks fingers or hands") were less directly interactional than others (e.g., "licks objects").

Reliability   A second individual was trained in the use of the coding manual and recoded 5 of the 11 videos. Intercoder reliability was established for all the categories.

How the Children Interacted with AIBO and Kasha

For all 11 children, the session with AIBO (M = 15.4 minutes) was longer than the session with Kasha (M = 10.1 minutes). This finding may be of interest in itself (e.g., it may suggest that children found AIBO more engaging than Kasha), but it is difficult to interpret with confidence as it could be due to other factors (e.g., experimenter bias). Accordingly, for all relevant statistical tests we accounted for this effect by focusing on the number of occurrences per minute (of the behavior of interest) rather than the total number of occurrences. In this way, our results can be considered conservative, with stronger effects almost always emerging were we not to make this adjustment.

Percentage of Interaction
We coded for how much time the child spent not interacting and inter-acting with each artifact during the session (see figure 9.2). The results

Figure 9.2
The same participant in "noninteraction" with Kasha (top) and in "authentic interaction" with AIBO (bottom).

were suggestive, but not quite statistically significant. Children spent an average of 72 percent of the AIBO session interacting with AIBO, and only 52 percent of the Kasha session interacting with Kasha.

Spoken Communication to Artifact
We examined the total number of words the children spoke to each of the artifacts. Results showed that the children spoke more words per minute to AIBO (M = 2.73 words) than to Kasha (M = 1.07 words).

Table 9.1
Occurrences of Behaviors Typical of Children without Autism by Artifact

Behavioral interaction category	Total occurrences (across all participants)		Occurrences per minute for each participant			
			Mean		Median	
	AIBO	Kasha	AIBO	Kasha	AIBO	Kasha
1. Verbal engagement	27	2	.180	.034	.123	0
2. Affection	93	61	.598	.694	.542	.260
3. Animating artifact	6	42	.037	.272	0	0
4. Reciprocal interaction	169	53	.970	.458	.757	.489
5. Authentic interaction	17	2	.105	.011	.091	0

Behavioral Interactions Typical of Children without Autism
We coded for five categories of behavioral interactions typical of children without autism: verbal engagement, affection, animating artifact, reciprocal interaction, and authentic interaction. As shown in table 9.1, children engaged in more verbal engagement, reciprocal interaction, and authentic interaction with AIBO compared to Kasha. The category of authentic interaction was broken down into two subcategories: dyadic (between child and artifact) and triadic (between child, artifact, and experimenter). Results showed both more authentic dyadic interaction and more authentic triadic interaction in the AIBO condition than in the Kasha condition. There was no difference in the amount of affection or animation that the children accorded AIBO or Kasha. (Note that it might appear from the total occurrences that children animated AIBO less than Kasha, 6 compared to 42 total occurrences, respectively. But this difference was driven by large amounts of animation with Kasha by just three participants, and was not statistically significant.)

Behavioral Interactions Typical of Children with Autism
We coded for 16 behaviors typical of children with autism: (1) rocks back and forth, (2) flicks fingers or hands, (3) high-pitched noise, (4) unintelligible sounds, (5) repeats words, (6) lines up objects, (7) prances/walks on tiptoe, (8) whirls/turns in circles, (9) inappropriate

pronouns, (10) uses third person for self, (11) withdraws/standoffish, (12) unreasonable fear, (13) licks objects, (14) smells objects, (15) lunging/darting, and (16) self-injurious. Results showed no statistically significant differences between AIBO and Kasha in the number of occurrences per minute of any of the individual behaviors. However, when we combined all the behaviors together, results barely missed statistical significance: the mean number of autistic behaviors per minute with AIBO was 0.75, and the mean number of autistic behaviors per minute with Kasha was 1.1.

Conclusion

The results from this study suggest that robot dogs can aid in the social development of children with autism. My colleagues and I found that in comparison to a mechanical toy dog (which had no ability to detect or respond to its physical or social environment) the children spoke more words to AIBO, and more often engaged in three types of behavior with AIBO typical of children without autism: verbal engagement, reciprocal interaction, and authentic interaction. In addition, we found suggestive evidence that the children interacted more with AIBO, and, while in the AIBO session, engaged in fewer autistic behaviors.

We utilized a new behavioral category in this study: "authentic interaction." We had first conceptualized this category by drawing on the last of nine benchmarks that my colleagues and I had proposed for human–robot interaction: the authenticity of relation (Kahn et al. 2007). At that time, we had defined this benchmark using Buber's (1996) account of an I–You relationship, wherein an individual relates to another with his or her whole being, freely, fully in the present, unburdened by conceptual knowledge. In turn, given our population of children with autism, we recast this category in somewhat less ethereal terms, and sought to capture what many of us probably take for granted in our social relationships. Namely, in relation to a social partner, we make nuanced adjustments in our facial expressions, body positions, and speech to respond to the other appropriately in context. That was partly how we defined authentic interaction in this study. Quantitatively, that was what we found. More children interacted authentically with AIBO than with Kasha.

To convey this new category illustratively, consider the following representative event from our data. (Figure 9.2, bottom, is a still photo of this event while it was in progress.) The child sits on the floor with AIBO and the experimenter. The child is trying to convince AIBO to eat the dog biscuit. The child's hand movements are gentle, his body position relaxed. He softly says to AIBO: "Open wide, here comes the choo-choo." The experimenter then asks the child if his mother did that with him when he was little, and he giggles and says no, and then adds, "I saw it on *Viva Piñata*." This interaction feels "authentic" insofar as it feels whole, engaged, nuanced, even intimate. The child is integrating gentle behavior with AIBO and appropriate imaginary speech ("Open wide, here comes the choo-choo") with an on-topic conversation with the experimenter, drawing on his past experience (of watching *Viva Piñata*). Contrast that interaction with the same child's interactions with Kasha, in which he looks disengaged (figure 9.2, top), and roams around the room and responds distractedly to the experimenter's questions, with answers such as "because she is," "yes," "yeah," and "uh-huh."

If animal robots can aid in the social development of children with autism, how does it happen? Perhaps it happens in one of two ways. One way is through the child engaging with the robot as a social other. This proposition was supported by our findings based on children's speech production, verbal engagement, reciprocal interaction, and authentic interaction. Granted, it remains an open question whether children's social engagement with a robot transfers to their social engagement with other people (Robins, Dautenhahn, and Dubowski 2005). Yet, at least for some forms of functioning, it seems to me that it would. For example, if (as in the current study) children with autism produce more coherent speech with a robot than with other play artifacts, such production would seemingly provide the children with increased capability that they could later utilize in human–human interaction. A second way is that robots might provide children with autism with a pivotal medium for enhanced communication with adults. We found support for this proposition insofar as children more often engaged in authentic interaction with the experimenter in the AIBO condition than in the Kasha condition.

Three limitations of this study are worth noting and responding to. One is that AIBO differed from Kasha in both form and function, and experimentally we were unable to disambiguate the two. Future studies would profit by moving in this direction. For example, one approach would be to compare an interactive AIBO with a noninteractive AIBO. That said, AIBO's programming does not allow for adding or subtracting behavioral units. Thus, this approach would require, in effect, the entire reprogramming of an AIBO: a substantial technical undertaking. A second limitation involves the small sample size (11 participants). Recruiting children for this study was much harder than in other studies we have conducted with typically developing children. The upshot was that we had little power in our statistical tests. Nonetheless, we attained statistical significance on the above key measures. A third limitation is that we did not compare AIBO to a biologically live dog, as was done in the study I describe in chapter 8. Had we done so, I suspect we would have found that the biologically live dog was more effective than AIBO in aiding children with autism. But even if that turns out to be true, the results from this study show that this form of technological nature may hold some benefits in future years for certain populations of children.

One of the symptoms of autism, as described earlier, is that children engage in repetitive behaviors. But based on our results, I wonder if the autism field has somewhat misunderstood this symptom. Granted, some of the children in our study enjoyed Kasha's repetitive behaviors, and they often engaged Kasha in their own repetitive behaviors. For example, one child lined Kasha up on the floor and then Kasha would walk in a straight line, and then the child would pick Kasha up and take Kasha back to the starting point, and the child repeated this behavior many times. But I think many of these children recognized something of what I recognized, that Kasha's repetitive behavior had a dreary sameness to it, compared to AIBO's behavior. Kasha made the exact same sound and moved in the exact same way every single minute it was turned on. In contrast, AIBO's repetitive behavior varied. For example, while AIBO often searched for its pink ball (which can be categorized as repetitive behavior), one never knew when it would do so, and when it did so the pattern varied: the ball was in a new location, AIBO approached it in different ways, oriented itself to the ball in different ways, and was more or less successful with its kick or head-butt depending on the many nuanced aspects of its programming. This difference between artifacts may partly explain why the children spoke more words to AIBO and engaged with AIBO with more verbal interaction, reciprocal interaction, and authentic interaction.

The key idea, then, is this: AIBO is patterned simply, but it is not simply repetitive. In this sense, AIBO is similar to much of nature. People can be captivated, for example, by looking at ocean waves breaking onto the shore, or experiencing a waterfall, because the water's pattern is always changing, simple but compelling. I suspect that these engaging "repetitive" properties of nature—and of technological nature if it is designed right—have not yet been fully recognized as a means to benefit children with autism.

10

The Telegarden

This book is focused on the psychological effects of interacting with technological nature. My primary empirical data has been based so far on six studies across two forms of technological nature: two studies on the display of real-time nature on flat digital screens (chapters 4–5) and four studies of people interacting with robotic pets (chapters 6–9). Each study took on the order of several years to complete. It is not quick research. Thus one needs to choose one's studies carefully. The advantage to multiple studies on a single form of technological nature is that we can gain confidence in the generalizability of the results for that form of technological nature. The advantage to investigating multiple forms of technological nature is that we can gain confidence in generalizing from any specific instantiation of technological nature to the phenomenon itself.

In this final empirical chapter, I provide data on a third form of technological nature, on what is called the Telegarden—an installation that allowed people to plant and tend seeds in a distant garden by controlling a robotic arm through a Web-based interface.

The Telegarden was developed at the University of Southern California in 1995 under the codirection of Ken Goldberg and Joseph Santarromana. In 1996 the Telegarden was moved to the Ars Electronica Center in Linz, Austria, where it remained online until August 2004 (Goldberg 2000; "The Telegarden Website," n.d.). The garden itself was a small plot encircling an industrial robotic arm. A Web-based interface allowed distant people to activate the robotic arm, view the garden through a camera mounted on the robotic arm, change the view, plant a seed, water it, and (if one was a successful gardener) water the resulting plant on an ongoing basis. Many thousands of distant people interacted with the Telegarden in such ways. In addition, the designers of the Telegarden sought to foster a sense of community (not unlike a community garden) by helping people become aware of each other and of the impacts of their actions on the garden and other people. Toward this end, the site provided information about who was currently visiting the site, what they were doing, and what actions people had taken in the garden recently. In addition, the site had a corresponding chat room, where people could engage about topics relevant to the garden, and any other topic of choice.

The purpose of this study was to investigate people's experience of the Telegarden. To do so, my colleagues and I conducted a line-by-line analysis of three months of chat discourse from the chat room associated with the Telegarden (Kahn, Friedman, Alexander, Freier, and Collett 2005). Our methods built on those we had used successfully in analyzing the AIBO online postings discussed in chapter 4. It was our thought that people's experience with the Telegarden could be quite rich, and could embody some of the human experience of gardening itself. For example, Schultz in *Garden Design* had written: "Though drained of sensory cues, planting that distant seed still stirs anticipation, protectiveness, and nurturing. The unmistakable vibration of the garden pulses and pulls, even through a modem" (as quoted in "The Telegarden Website," n.d., para. 6). Yet it remained an open empirical question whether such writing conveyed more of the writer's exuberance than fact in terms of people's experience of the Telegarden.

Four overarching questions structured our investigation. First, to what extent did the Telegarden actually engage people in nature-oriented experiences? Second, were such experiences grounded by the physicality of the Telegarden itself, or largely focused on nature beyond the Telegarden, in, for example, the physical location of each person? Third, how did people experience the technological aspects of the installation, particularly the robotic arm that was the mechanism for acting in the garden? Fourth, did people's experience of the Telegarden change, and perhaps deepen, with increased participation?

A three-month (13-week) period, comprising 22,952 posts, was selected for coding. These three months represent a period of robust use and exploration of the Telegarden site, soon after the installation was relocated to Austria. From this group of 22,952 posts, we then excluded all posts that were written in a non-English language or that were exact repeat posts (which was common due to the slow response of the installation's server). The resulting data source for analysis comprised 16,504 posts. Our coding system was first generated from postings in the archive outside the three-month period selected for coding. Once finalized, the coding manual was used to code the three-month data set. If a single post included several instances of a single category, that category was coded as "used" only once. It was also possible for several codes to apply to a single post, to account for the expression of a range of topics and attitudes within one statement. To assess reliability of the coding system, an independent scorer trained in the use of the coding system recoded 15 percent of the total postings across the three-month period. Good intercoder reliability was established for the coding system. Our full coding system is available as a technical report (Kahn, Friedman, and Alexander 2005).

Conversational Categories in the Telegarden

As shown in table 10.1, three overarching categories of conversation in the Telegarden characterized most of the chat discourse:

Table 10.1
Percentage of Postings by Conversational Category (Total Postings – 16,504; Number of Members Posting = 347)

Conversational Category	Percentage of Postings	
1. Conversations about Nature	13	
1.1 Experience of garden within Telegarden	6	
1.1.1 Geography		*1*
1.1.2 Watering		*2*
1.1.3 Planting		*2*
1.1.4 Overcrowding		*1*
1.1.5 Other		*2*
1.2 Nature beyond the Telegarden	7	
1.2.1 Gardening in general		*0*
1.2.2 Weather		*3*
1.2.3 Seasons		*1*
1.2.4 Environmental issues		*1*
1.2.5 Other		*3*
1.3 Other	0	

Table 10.1
(continued)

Conversational Category	Percentage of Postings		
2. Conversations about Technology	**22**		
2.1 Experience of technology within Telegarden		9	
2.1.1 Robot			*0*
2.1.1.1 Artifactual description				*0*
2.1.1.2 Biological description				*0*
2.1.1.3 Agency				*0*
2.1.1.4 Social standing				*0*
2.1.1.5 Moral standing				*0*
2.1.1.6 Other				*0*
2.1.2 Interface widgets and gizmos			*5*
2.1.3 Interaction design			*1*
2.1.4 AEC staff intervention			*1*
2.1.5 Other			*3*
2.2 Technology beyond the Telegarden		13	
2.3 Other		0	
3. Conversations about Humans	**69**		
3.1 Experience of community within Telegarden		13	
3.1.1 Friendship and interpersonal support			*2*
3.1.2 Inclusion/welcoming/new user			*2*
3.1.3 Origin stories/history			*2*
3.1.4 The chat room and chat			*5*
3.1.5 Other			*3*
3.2 Social life beyond the Telegarden		58	
3.2.1 Identification			*1*
3.2.2 Friendship beyond the Telegarden			*1*
3.2.3 Family			*2*
3.2.4 Work			*3*
3.2.5 Travel/vacation			*2*
3.2.6 Chit-chat			*45*
3.2.7 Other			*8*
3.3 Other		0	
4. Uncodable	**7**		

Note. The percentages reported in **bold** refer to usage of the overarching category; percentages in plain text refer to the next sublevel in the hierarchy; and percentages in *italic* refer to the lower levels. Within each level of the hierarchy, postings that contained more than one subcategory are only counted once in the overarching category.

1. Conversations about Nature (13%) referred to people's discussion of actual nature (often connected to gardening) within and beyond the Telegarden. As people wrote: "It's fun to sometimes catch site of a lady bug crawling about in the soil"; "What is cool is that we are moving and I can take my garden with me!" We coded for but found no (0%) evidence of dialogue that plants deserved care or respect, or statements of people talking to their plants. *2. Conversations about Technology* (22%) referred to people's discussion of technology within and beyond the Telegarden. As people wrote: "The image to the left is one you use to shift the robot . . . you can tell if the space is occupied by the green dots"; "Aren't you just a bit amazed you can actually control a robot in a foreign country through the use of a computer from the comfort of your own home?" We coded for but found no (0%) evidence of dialogue about the biological, mental, social, or moral standing of the robotic arm (or robotic installation) itself. *3. Conversations about Humans* (69%) referred to people's discussion of an interpersonal connection within and beyond the Telegarden. As one person wrote: "She's such a lovely girl, now that I've met her I know I cannot be without her, she means so much."

Since one of our central questions in this study focused on whether people experienced the nature part of telegardening in a deep and meaningful way, I think it would be helpful to look at the specific subcategories and examples of the above general category "Conversations about Nature." There were four specific subcategories: geography, watering, planting, and overcrowding.

The geography category involved dialogue about (a) where the Telegarden is, (b) its size, (c) specific locations of plants in the Telegarden, and (d) other questions or comments about specific physical features of the garden or the history or design of the garden. Examples include the following (punctuation, spelling, and grammar are copied verbatim from original chat developed in the coding manual):

I AM COMPLETELY LOST AND NEED SOMEONE TO TALK TO ABOUT THIS WHOLE GARDEN THING THATS GOING ON.I MEAN,JUST WHER IS THE GARDEN AT ?

I read there are over 6000 plants in there. It must be a huge robot garden. How could there be that many?

The garden has more plants than ever before-don't you think?

Look at all the pretty flowers at I5.. No, we're not on the freeway..

what is this plant at sector p8

The watering category involved (a) watering as an individual task, (b) watering as a communal task, (c) watering as enjoyable for its own sake, (d) watering as beneficial to plants and the garden, and (e) watering as destructive to plants and the garden. Examples include the following:

My watering is done.

I am at a telecomputing workshop. Has anyone watered plants today?

Nice watering...one day I was here alone and considered watering a large smile face,but figured people would think I was an idiot:)

Hey Michel, where is your seed planted? I want to go water it!

Vince, easy on the water your flooding my seed out. Spread it around.

The planting category involved (a) plans for planting, (b) types of seeds to plant, (c) finding a clear spot to plant, (d) planting after the garden is cleared out, and (e) tending and caring for plants or seeds planted. Examples include the following:

Hold on a sec, let me plant one real quick

Too bad we can't plant flowers or food—like strawberries

Sylvia, there seems to be a pretty good bare spot @12.96 and 3.68

The garden is scheduled to be cleared out in early May to start anew, I Have a seed to plant, but will wait for the new garden.

Take care All I think plant is doing as well as it can right now.:-)

The overcrowding category involved (a) overcrowding in the garden and (b) clearing of the garden. Examples include the following:

no room anyway...can 10263 plants grow in such a small garden

the TG looks like a jungle

when they replant the garden, will the seeds I just planted remain?

They cleared it once before and i think i just got my old placings back. I don't remember if i got new seeds.

We coded for insulting or abusive language (flaming) but found virtually none (0.1%, 21 out of 16,504 postings, with 20 out of the 21 postings coming from a single person).

Some other events that struck us when reading through the chat discourse include the following: People sometimes asked others to water their plants while they (the initiating people) were on vacation (and presumably off-line). People on occasion planted strategically to be nearby specific other people. The weather in a person's own location was commonly shared in conversations. There was a death of a person using the Telegarden, and the community responded by diffusing the information to the community and expressing sorrow. There was an engagement in the Telegarden community, and the couple maintained the Telegarden as their meeting ground even though they often would use other means to communicate. The Telegarden also provided a means for people to garden who physically could not tend to a real garden. As one person wrote: "I am recovering from neck surgery and can not do anything. The place has been a life saver for me . . . almost like meeting in a community garden!! A community of world gardeners, I guess!!"

Conversational Response to Physical Activity in the Telegarden

One striking feature of telerobotics in general, and the Telegarden in particular, is that one acts remotely in a real physical world (as opposed to a virtual world). Thus, we tried to be alert to how conversations shifted and coalesced based on real events in the garden itself. For example, on occasion, the Telegarden would become too full of plants, and the garden itself was physically cleared by the Telegarden administrators, to allow room for new plantings. One such clearing occurred during this three-month coding period, at the end of week 5. Immediately after the clearing of the garden, conversations tripled (from 3% to 9%) concerning the experience of the Telegarden as garden (category 1.1 in table 10.1). For example, one person said: "Hey it works . . . planted my seed!!! I am really glad to see . . . the 'new'-new beginning of the garden . . . I'm sure I'll get to see alot of the oldtimers drop by now . . . News travels rather fast in these parts."

Differences in Conversations with Increased Participation

To investigate how people's conversations varied with increased participation in the Telegarden, we segmented people into quartiles (four groups as evenly as possible based on the number of postings per person). As shown in table 10.2, as individual participation in the chat room increased, conversation decreased about nature and technology within the Telegarden (e.g., "there is sure a pretty flower"), and increased about nature and technology beyond the Telegarden (e.g., "seems everything is blamed on El Nino now"). This finding (which was statistically significant) might make it seem that, with increased participation, the Telegarden itself—the actual garden and installation—engaged people psychologically less and less. But it seemed to us that what was actually happening was that the less-involved people (those in the first and second quartiles who posted less than a total of 7 times) were unduly focused on getting up to speed with planting and learning how to use the technology, thereby muddling the picture of what engaged participation with the Telegarden looked like.

Thus, we conducted an additional analysis focused on only the people in the fourth quartile (86 people total), those who posted at least 22 times. We rank-ordered these participants based on their percentage of overall posts on technology and nature. We then plotted their respective percentages of posts within the Telegarden and beyond the Telegarden. Next, we overlaid the scatter plots with linear regression lines. Based on visual inspection of the resulting data, the findings showed us the following. The more a person's conversations were about technology, the more likely it was that the person talked about technology beyond the Telegarden versus within the Telegarden. In contrast—and this is the striking finding—the more a person's conversations were about nature, the more likely it was that the person talked about nature within the Telegarden versus nature beyond the Telegarden.

Conclusion

Telerobotics is now a part of our lives. Surgeons, for example, use telerobots to conduct surgery from distant locations. Oceanographers use telerobots to explore the depths of the oceans. NASA has roamed the surface of Mars using a telerobot. On the military side, telewarfare is increasingly emerging. Unmanned aerial vehicles, for example, fly through enemy airspace for the purpose of reconnaissance and attack, and are often controlled directly from the ground in a distant location.

Table 10.2
Percentage of Postings for Each Quartile by Conversational Category

Conversational Category	1st Quartile R = {1–2} N = 110 P = 140	2nd Quartile R = {3–6} N = 74 P = 313	3rd Quartile R = {7–21} N = 77 P = 941	4th Quartile R = {22+} N = 86 P = 15110
1. Conversations about Nature	**20**	**17**	**18**	**15**
1.1 Experience of garden within Telegarden	16	13	13	8
1.3 Other	1	1	1	0
2. Conversations about Technology	**9**	**25**	**21**	**21**
2.1 Experience of technology within Telegarden	5	19	14	10
2.2 Technology beyond the Telegarden	4	5	7	12
2.3 Other	0	1	0	0
3. Conversations about Humans	**72**	**63**	**70**	**69**
3.1 Experience of community within Telegarden	20	16	17	13
3.2 Social life beyond the Telegarden	52	49	55	57
3.3 Other	0	0	0	0
4. Uncodable	**7**	**7**	**6**	**6**

Note. (1) R = range of postings per member; P = number of postings per member; N = number of people; P = number of postings per quartile. (2) Quartiles for each of the four groups were established as evenly as possible based on the number of postings per member. (3) When postings contained material from more than one category, all categories were coded; thus, the sum of the percentages of postings across categories may be greater than 100. (4) The percentages reported in **bold** refer to usage of the overarching category; percentages in plain text refer to the next sublevel in the hierarchy. Within each level of the hierarchy, postings that contained more than one subcategory are only counted once in the overarching category.

In the coming years, these telerobots, and others, will become more sophisticated in terms of their computation and hardware, and more pervasive on a societal level. What is novel about a Telegarden is that it applies telerobotics not to surgery, geological or planetary exploration, or the agendas of warfaring nations, but to human–nature interaction. A Telegarden seeks to encourage and make possible the experience of gardening.

Given this context, we sought to understand people's experience and conceptions of the specific Telegarden created by Ken Goldberg, Joseph Santarromana, and their colleagues. How well did this Telegarden work as a means to connect people to nature?

Our results suggest a simple answer: "not so well." Quantitatively, only 6 percent of 16,504 posts referred to people's discussion of the nature within the Telegarden; and only an additional 7 percent referred to people's discussion of nature beyond the Telegarden. The reader might want to pause here and go back and reread the 19 representative examples I provided earlier of the chat about nature within the Telegarden. As I read this qualitative data, the chat discourse about nature seems "thin." People did not, for example, talk about the plants in the Telegarden in biocentric terms of deserving care or respect. People did not talk about talking to their plants. Perhaps that is not surprising, because the plants were in a distant location and would not hear one's vocalized speech. Granted, on rare occasions a post would strike me as comparatively deep and important, as in the one quoted earlier: "I am recovering from neck surgery and can not do anything. The place has been a life saver for me . . . almost like meeting in a community garden!!" But that was unusual. What then do we make of the characterization of the Telegarden, quoted earlier, from *Garden Design*: "Though drained of sensory cues, planting that distant seed still stirs anticipation, protectiveness, and nurturing. The unmistakable vibration of the garden pulses and pulls, even through a modem"? In my view, that characterization is more hyperbole than fact.

A difficult question emerges, however, of how to understand the paucity of human–nature connection in the Telegarden. The Telegarden was the first of its kind, using rudimentary technologies with slow response time and limited capabilities. The visual interface was also poor, providing little sense of the garden as a whole, of its texture or configuration, or of individual plants themselves. In this light, the 13 percent figure (the percentage of posts focused on nature) could be seen as quite large. We also found from the fourth quartile group (who posted at least 22 times) that the more a person's conversations were about nature, the more likely it was that the person talked about nature within the Telegarden versus nature beyond the Telegarden. This latter finding provides evidence that at least some active people (those drawn most to nature in general) sustained their interaction with nature as embodied within the Telegarden.

To put these findings in perspective, imagine living back in 1926, and watching "nature movies" for the first time using John Baird's television that operated with 30 lines of resolution at 5 frames per second. That technologically mediated experience with nature would be quite limited compared to today's standard. Thus, there is reason to believe that as the telerobotic technology advances, the connection between people and natural places will become increasingly compelling. But as I have asked across this book's empirical chapters: how compelling?

The answer is, I believe, the same answer that I have offered for the other forms of technical nature discussed in this book. As with visually experiencing real-time nature on flat digital screens (chapters 4–5) and interacting with a robot pet (chapters 6–9), in the years to come gardening in a Telegarden will likely be better than experiencing no nature but not as good as experiencing actual nature.

11
Environmental Generational Amnesia

I started recognizing the problem of environmental generational amnesia some years ago. A colleague and I had interviewed African-American children in the inner city of Houston, Texas, about their environmental views and values (Kahn and Friedman 1995). We talked with the children about dozens of issues, such as whether animals, plants, and open spaces played a role in their lives, and if so how, and whether it was all right or not to throw trash in their local bayou, and why, and whether their judgments generalized to a people elsewhere with different environmental practices. We found that these children—72 of them across grades 1, 3, and 5—spoke of the importance, personally and morally, of environmental issues in their lives. Often their reasons were what we termed *anthropocentric*, meaning that the children focused on protecting nature so as to advance human goals. One child said, for example, that it was not all right to pollute the air because "air pollution goes by and people get sick, it really bothers me because that could be another person's life." Occasionally their reasons were what we termed *biocentric*, meaning that the children believed that nature had intrinsic value or rights. One child said, for example, that it was not all right to pollute the local bayou because "water is what nature made; nature didn't make water to be purple and stuff like that, just one color. When you're dealing with what nature made, you need not destroy it."

At the time, there was a stereotype held by many in the United States that African Americans, especially those living in urban poverty, were not interested in nature or environmental issues. Most of our research findings spoke against this stereotype. Those findings did not surprise us. The desire for and connection with nature runs deep in the human psyche, as I discussed in chapters 1 and 2.

But one set of findings did surprise us. We had asked the children whether they knew about air pollution and water pollution. Two-thirds of the children did, and could often explain the constructs well. Then we asked these same children whether they thought that Houston had air pollution or water pollution. We found that only one-third of the children thought so, a statistically significant difference. That difference surprised us because Houston is one of the most environmentally polluted cities in the United States. Local oil refineries contribute not only to the city's air pollution but also to distinct oil smells during many of the days. Bayous can be thought of more as sewage transportation channels than fresh water rivers. Thus we had asked ourselves, how could it be that children who knew about pollution in general, and who grew up in a polluted city, not be vividly aware of their own city's pollution? Our answer was that to understand the idea of pollution one needs to compare existing polluted states to those that are less polluted. These children in Houston had little experience with less polluted environments for such a comparison.

I have seen first-hand, over several decades, a similar psychological process enacted on mountain land in Northern California, though tied not to air or water pollution but to the health and integrity of the forests. Here is a synopsis of what I have seen. A family moves to a piece of forested land, say, 640 acres, a square mile, which has already been logged numerous times in the last century. These are usually good people. They might well view themselves as environmentalists. They might be members of the Sierra Club. But like most of us, they need to make ends meet, and so they look around at the natural resources, the timber, and they say: "Well, there should be a way of taking some timber here, and still leave some good trees. You know, all of us use wood products, so it's kind of hypocritical to be saying no logging." So they log. Then they say: "You know, 640 acres, what are we really going to do with that much land? And if we sell some, then we can make our land payments." So they subdivide the land into four 160-acre parcels, keeping the nicest parcel for themselves. Families from more urban areas now buy each of the remaining 160-acre parcels. These, too, are usually good people, even environmentalists. And they say something like: "Well, there should be a way of taking some timber here, and still leave some good trees. You know, all of us use wood products, so it's kind of hypocritical. . . ." So these families log the land, and afterward subdivide into 40-acre parcels, if the zoning laws allow. Notice how relative is the concept of "good." Each logging and subdivision degrades the land more, but each person assesses the health and integrity of the land relative to a more environmentally degraded urban setting, and not to the land's condition as it was even a year before.

The children in Houston grew up with an existing level of pollution, and that amount became for them the normal, nonpolluted state. The adults in the above scenario had conceptualized the health and integrity of mountain forests from the vantage point of their urban setting. In the same way, all of us construct a conception of what is environmentally normal based on the natural world we encounter in childhood. The crux is that with each ensuing generation, the amount of environmental degradation can and usually does increase, but each generation tends to take that degraded condition as the nondegraded condition, as the normal experience. It is a condition that I have called *environmental generational amnesia* (Kahn 1999, 2002; Kahn, Severson, and Ruckert 2009).

I think environmental generational amnesia is an enormous problem that exists worldwide. I will say more about the problem shortly. But I recognize that the evidence for it has not been as strong as it should be, which allows people to ignore the problem further. As an example, the National Park Service commissioned a report entitled *A Critical Review of the Concepts of "Environmental Generational Amnesia" and "Nature Deficit Disorder"* (Swanson, Johnson, and Kamp 2006). The latter is a term Richard Louv (2005) uses in his widely read book, *Last Child in the Woods*. The Park Service asked me to respond to their critique of environmental generational amnesia. Their critique was that I did not have strong enough scientific evidence. In my response I said that that was true. However, I also noted that in their document title they say that they are reviewing the "concept" of environmental generational amnesia. But in their review they only examined its empirical base. I also reminded them that for more than twenty years the U.S. government said that there was not enough scientific evidence to substantiate the hypothesis about global warming. I pleaded with them to take a leadership role in enhancing the human relationship—both domestic and wild—with the wonderful park lands that are within their trust. My words fell short.

As a scientist, I still do not have the evidence that I would like to substantiate environmental generational amnesia. Let me then be a little more careful than I was in years past. Let me call it a hypothesis. My hypothesis is that environmental generational amnesia exists. With that casting, in this chapter I would like to make this hypothesis compelling to the reader. I will first provide some historical and current snapshots of it, and then provide further evidence in its support. Then, assuming that the reader believes that there may be some merit to this hypothesis, I discuss possible solutions.

Snapshots Past and Present of Environmental Generational Amnesia

It is not uncommon to read accounts, current and historical, of experiences that people have and have had with nature that highlight environmental generational amnesia. Here are some examples.

The Highlands of Scotland

Many centuries ago the forests in the Highlands of Scotland flourished. According to Hand (1997) these forests were "grand as any on earth. Elm, ash, alder, and oak shaded the low-lying coastal plains and inland valleys; aspen, hazel, birch, rowan, and willow covered the hills; and beautiful, redbark Scots pine clung to the glacial moraines and steep granite slopes. The Romans called it the Forest of Caledonia, 'the woods on heights,' and it clung to Scottish soil for millennia" (p. 11). However, at the start of the sixteenth century, with the coming of the English and the industrial revolution, the forests came under siege, and by the 1700s had been virtually eliminated.

Stone houses and coal fires replaced those of wood. Soils, exposed to harsh winds and rain, washed into streams and rivers, leaching fertility, destroying fisheries. Erosion cut, in many places, to bedrock. Woodland species—bear, reindeer, elk, moose, beaver, wild boar, wild ox, wolf (the last killed in 1743), crane, bittern, great auk, goshawk, kite, and seaeagle—vanished. . . . By 1773, when Dr. Samuel Johnson toured the highlands, with James Boswell, the landscape was, in Johnson's words, a "wide extent of hopeless sterility." He remarked that one was as likely to see trees in Scotland as horses in Venice. (Hand 1997, p. 12)

Today the Highlands of Scotland are one of the most deforested lands in the world. Perhaps equally disturbing, the Scots of today, according to Hand, have virtually no conception of a forest, of its ecological vastness and beauty. Hand presented these ideas in an essay he titled *The Forest of Forgetting*. It is a forgetting that has crossed generations.

Passenger Pigeons

None of us living today has experienced certain forms of interaction with nature that were common even one or two hundred years ago. For example, John Muir (Teale 1976) wrote of experiencing the immense migration of the passenger pigeons: "I have seen flocks streaming south in the fall so large that they were flowing over from horizon to horizon in an almost continuous stream all day long, at the rate of forty or fifty miles an hour, like a mighty river in the sky widening, contracting, descending like falls and cataracts, and rising suddenly here and there in huge ragged masses like high-plashing spray" (p. 46). Similarly, in the early 1800s, John Audubon wrote:

The air was literally filled with Pigeons; the light of noon-day was obscured as by an eclipse. . . . I cannot describe to you the extreme beauty of their aerial evolutions, when a Hawk chanced to press upon the rear of a flock. At once, like a torrent, and with a noise like thunder, they rushed into a compact mass, pressing upon each other towards the centre. In these almost solid masses, they darted forward in undulating and angular lines, descended and swept close over the earth with inconceivable velocity, mounted perpendicularly so as to resemble a vast column, and, when high, were seen wheeling and twisting within their continued lines, which then resembled the coils of a gigantic serpent. Before sunset I reached Louisville, distant from Hardensburgh fifty-five miles. The Pigeons were still passing in undiminished numbers, and continued to do so for three days in succession. The people were all in arms. The banks of the Ohio were crowded with men and boys, incessantly shooting at the pilgrims, which there flew lower as they passed the river. Multitudes were thus destroyed. For a week or more, the population fed on no other flesh than that of Pigeons, and talked of nothing but Pigeons. (American Museum of Natural History 2008)

It is hard to believe, but humans wiped out the passenger pigeon. The last one died in 1914 in the Cincinnati Zoological Garden. It is now an extinct species.

The North American Buffalo

We wiped out most of the American buffalo, as well. Lewis and Clark wrote of this experience in their journal (their original spelling and punctuation from the early 1800s are left intact):

I sent the hunters down Medicine river to hunt Elk and proceeded with the party across the plain to the white bear Islands. . . . it is now the season at which the buffaloe begin to coppelate and the bulls keep a tremendious roaring we could hear them for many miles and there are such numbers of them that there is one continual roar. our horses had not been acquainted with the buffaloe they appeared much allarmed at their appearance and bellowing. when I arrived in sight of the whitebear Islands the missouri bottoms on both sides of the river were crouded with buffaloe. I sincerely belief that there were not less than 10 thousand buffaloe within a circle of 2 miles arround that place. (Moulton 1993, pp. 104–106)

In the same way we do not experience the plethora of boundless winged life of the passenger pigeons, so do we not experience the plethora of the plain buffaloes, or the plethora of virtually any animal. We might think we do. But that is because we do not know what we are missing. We have lost those experiences.

Pyle's Educational Activity

Robert Michael Pyle, nature-writer and butterfly expert, often speaks to groups about nature, and during his speaking it is not uncommon for him to engage an audience in a brief activity (Pyle 2002). He asks the audience members whether they can remember a particular place from their childhood in nature where they would go and play and explore with friends, or perhaps a place to be by themselves, or a place where they would make forts, catch bugs, or be in the water. Usually most of the hands would go up in the audience. He would then ask people to describe these places. Sometimes the places involved creeks or ponds, or a big tree, undeveloped parks, and old fields. Sometimes the places involved a vacant lot in the city: a place at once near, secretive, in some ways wild, and full of possibility. Pyle writes: "Most people can relate the details of the spot and tell stories from their places that surprise even themselves with their remarkable clarity and nuance and the deep affection aroused" (p. 306). Finally, Pyle asks a question that can lead audience members to feel some sadness: "How many can return to their special places and find them substantially intact?" (p. 306). Not many people can. The tree may have been cut down. The waterway may have been filled in or diverted, making room for a new shopping plaza. Condos might now have been built on the vacant lot, or a highway might speed through it.

Pyle's educational activity encourages us to recognize the environmental losses within our own lifespan; and it allows us to ask which of these losses our children care about, or even understand intellectually.

The Wilderness Society

Take a guess what year the following magazine editorial excerpt was written: "This [society] is born of an emergency in conservation which admits of no delay. It consists of persons distressed by the exceedingly swift passing of wilderness in a country which recently abounded in the richest and noblest of wilderness forms, the primitive, and who purpose to do all they can to safeguard what is left of it." In the last decade we have indeed witnessed the swift passing of wilderness in the United States; and environmentalists often speak of this problem as one which admits of no delay. The above passage was written, however, in 1935 as the opening to the first issue of the magazine for the Wilderness Society ("First Issue," 1993, p. 6). Thus environmental problems can be understood as equally serious across generations even while the problems worsen.

Mt. Whitney

Meloy (1997) writes that in 1929 her mother, then a child,

bellied up to the edge of a sheer cliff on a 14,495-foot Sierra peak and, while someone held her feet, stared down into empty blue-white space. Local newspapers reported her as the first child to climb Mt. Whitney. "On that three-week trip we saw one other pack train from a distance," [her mother] recalled, "and we said the mountains were getting crowded." . . . [Now] thirty million people live within a day's drive of Sequoia and Kings Canyon parks. Space atop Mt. Whitney is rationed: you need a reservation to climb it from the east. (pp. 4–5)

Yet people today still speak of such outings in Kings Canyon as "wilderness" outings; and "noncongested" can refer to a packed freeway in the middle of Los Angeles as long as the cars are moving along in a timely fashion. Apparently—as in the case of the above snapshot from the Wilderness Society—across generations the same environmental construct can refer to worsening environmental conditions.

Rowing to Latitude

Over several decades, Fredston (2001) rowed more than twenty thousand miles of some of the wildest coastlines in the arctic waters. During one of her later expeditions, she and her husband were rowing along portions of Norway. She describes some of the beauty of the land. But then she adds:

Still, even the undeniably beautiful portions of the Norwegian coast that send visitors from more developed, congested parts of Europe into raptures seemed sterile to us. . . . That experience frightened us to the marrow. It made us realize that, like the perpetually grazing sheep [in Norway], centuries of human habitation have nibbled away not only at the earth but at our perception of what constitutes nature. When we do not miss what is absent because we have never known it to be there, we will have lost our baseline for recognizing what is truly wild. (p. 217)

In that last sentence Fredston succinctly explains the problem of environmental generational amnesia as it applies to wildness.

Further Evidence for Environmental Generational Amnesia

The snapshots could go on and on. How many would it take—across geography and time—to make the hypothesis of environmental generational amnesia compelling to the skeptic? One hundred? Five hundred? Or would the skeptic always assert the dictum that "the plural of anecdote is not data"?

There is not too much experimental data that directly tests the hypothesis of environmental generational amnesia. The best study that I am aware of was conducted by Evans, Jacobs, and Frager (1982). They established two groups of participants. One group, long-term residents, had lived in the Los Angeles area five years or more. A second group, new arrivals, had just moved to the Los Angeles area within the previous three weeks. The groups were otherwise comparable in terms of socioeconomic status, age, gender, and previous health conditions. The Los Angeles area had at the time (and still has today) some of the worst air pollution in the United States. Of the four variables the researchers investigated, two are relevant for our purposes: what they called *problem assessment* (whether participants believed that Los Angeles had a problem with air pollution) and *signal detection* (whether participants could detect smog in photographs). In terms of problem assessment, first participants rank-ordered the five most negative aspects of living in their community, and second, estimated the amount of smog in their community on a 1–100 absolute-magnitude estimation scale. In terms of signal detection, participants viewed slides of different urban and rural Southern California scenes, which depicted differing levels of smog. Each slide was shown for 20 seconds, and participants were asked to judge whether or not smog was present. Results showed that long-term residents of the Los Angeles area, in comparison to the new arrivals, judged that smog was less of a problem in the area, and less often spontaneously mentioned smog. In terms of signal detection, both groups of participants were equally sensitive in detecting the presence of substantial amounts of smog in photographs; however, for low levels of smog, long-term residents in comparison to the new arrivals were less likely to report that smog was present. Taken together, these results offer evidence for environmental generational amnesia insofar as they support the proposition that people who live with a certain level of air pollution for an extended period of time become desensitized to that pollution, and less readily recognize that such pollution exists.

From the above study, we learn that people who move to a more environmentally degraded area are more likely to recognize the environmental degradation in that area than are that area's long-term residents. In turn—and consistent with the hypothesis of environmental generational amnesia—when the reverse migration occurs (when people move to less environmentally degraded areas), the pattern of recognizing environmental degradation also appears to reverse (in that these people are less likely to recognize the environmental degradation of the new area). For example, Papworth et al. (2009) conducted a study that focused on perceptions of bird population trends of inhabitants of a village in East Yorkshire, United Kingdom. Fifty inhabitants (3.4% of the village population) were asked (a) to identify five species of birds from photographs and to state the direction of change in each species' abundance between the year they moved to the village and now, and (b) to list the three most common birds in the village now and 20 years ago. Results showed that respondents who had lived elsewhere in Yorkshire or an urban area prior to moving to the village, compared to respondents who had presumably grown up in the village, were more likely to have a perception that there had been no change in bird population from 20 years ago. Of those respondents who perceived no change, they were also more likely to believe that the most common bird species now were also the most common bird species 20 years ago, which, based on count data from years past, was not the case. In other words, if we let bird population (both in number and kind) be a proxy for environmental conditions, this study shows that "recent arrivals" to the village were more likely than were long-term residents to believe the current environmental conditions in the village were the same now as they were 20 years ago.

Further evidence to support the hypothesis of environmental generational amnesia can be gained by assessing whether it provides explanatory power to historical events. I think it does. As a case in point, consider the human and environmental history of Easter Island (Diamond 2005; Flenley and Bahn 2003; Tilburg 1994). Humans arrived on Easter Island by 900 CE. They found a subtropical forest, whose species of trees are used elsewhere in Polynesia for making canoes, rope, tapa cloth, harpoons, and outriggers, and which yield edible fruit, are suitable for carving and construction, and are excellent for wood fires. One of those trees was the palm tree, whose trunk could exceed seven feet in diameter. As Diamond writes: "The trunk yields a sweet sap that can be fermented to make wine or boiled down to make honey or sugar. The nuts' oily kernels are rated a delicacy. The fronds are ideal for fabricating into house thatching, baskets, mats, and boat sails. And of course the stout trunks would have served to transport and erect moai [the 12-ton statues], and perhaps to make rafts" (p. 103). A thriving human population took hold on this island. Yet by 1722, when the Dutch explorer Jacob Roggeveen first landed on Easter Island, the land was completely deforested, with all of its tree species extinct. With the trees gone, the islanders had not been able to build stout canoes to fish the waters, wildlife had decreased, land birds had disappeared completely, wild fruits no longer grew, erosion had led to decreased crop yields, and there was no firewood with which to keep warm on the island's winter nights. People starved. There was a population crash. Cannibalism emerged. "Oral traditions of the islanders are obsessed with cannibalism; the most inflammatory taunt that could be snarled at an enemy was 'The flesh of your mother sticks between my teeth'" (ibid., p. 109).

How could the people of Easter Island cut bare their life-sustaining forests? Granted, Diamond (2005) discusses nine physical features of Easter Island that made it more susceptible to deforestation: it was a dry, small, low, remote, cold high-latitude, old volcanic island, without makatea (coral reef thrust into the air), far from volcanic aerial ash fallout, and far from Central Asia's dust plume. But even given the island's susceptibility, Diamond argues that it was "a society that destroyed itself by overexploiting its own resources" (p. 118). Thus the question still remains, how could they let that happen? Diamond's answer is as follows:

"What did the Easter Islander who cut down the last palm tree say as he was doing it?" We unconsciously imagine a sudden change: one year, the island still covered with a forest of tall palm trees being used to produce wine, fruit, and timber to transport and erect statues; the next year, just a single tree left, which an islander proceeds to fell in an act of incredibly self-damaging stupidity. Much more likely, though, the changes in forest cover from year to year would have been almost undetectable: yes, this year we cut down a few trees over there, but saplings are starting to grow back again here on this abandoned garden site. Only the oldest islanders, thinking back to their childhoods decades earlier, could have recognized a difference. Their children could no more have comprehended their parents' tales of a tall forest than my 17-year-old sons today can comprehend my wife's and my tales of what Los Angeles used to be like 40 years ago. Gradually, Easter Island's trees became fewer, smaller, and less important. At the time that the last fruit-bearing adult palm tree was cut, the species had long ago ceased to be of any economic significance. That left only smaller and smaller palm saplings to clear each year, along with other bushes and treelets. (p. 426)

Diamond (2005) calls this psychological phenomenon "landscape amnesia." It is people "forgetting how different the surrounding landscape looked 50 years ago, because the change from year to year has been so gradual" (p. 425). Diamond says that landscape amnesia is a "major reason why people may fail to notice a developing problem, until it is too late" (p. 426).

Landscape amnesia also helps account for the collapse of cultures in the U.S. Southwest, including "Mimbres around A.D. 1130; Chaco Canyon, North Black Mesa, and the Virgin Anasazi in the middle or late 12th century; around 1300, Mesa Verde and the Kayenta Anasazi; [and the] Mogollon around 1400" (Diamond 2005, p. 137). For example, when Native Americans initially moved into the Chaco Canyon area around 600 CE, they enjoyed a rich woodland of pinyon trees. These trees provided nuts for local food, timber for construction, and firewood. By 1000 CE the woodlands had become completely destroyed, and the land then became treeless. From there, Chaco residents proceeded to harvest ponderosa pine, spruce, and fir trees, up to 50 miles away. Diamond writes: "With no draft animals available, about 200,000 logs weighing each up to 700 pounds were carried down the mountains and over that distance to Chaco Canyon by human muscle power alone"

(p. 147). The last of those trees were stripped by about 1200 CE. Shortly thereafter Chaco Canyon collapsed environmentally and culturally. Again we can ask, as we did for Easter Island, how could a people cut their last tree? And after they experienced the loss of livelihood with the death of the pinyon trees, how could they then make the same mistake by deforesting the far-surrounding pines, spruces, and firs? The most parsimonious answer again lies with the amnesia hypothesis. It was not the same people making the same mistakes. Each generation likely started afresh constructing a new baseline for what was normal to their land-scape, and so as they continued to cut, they likely perceived their actions as causing relatively small harms, which could be justified based on the immediate benefits of the natural resource. They may have said what some of my neighbors have said in the mountains of Northern California, that we all use wood so it's a little hypocritical to be saying no logging. If they said that, it did not keep their society from collapsing.

While it is a only minor variation on terms, I prefer "environmental generational amnesia" over "landscape amnesia" because as I see it the major psychological mechanism that allows for enormous amounts of environmental devastation—enough to collapse a society—occurs not within a single person's life span, but across generations, through the construction of environmental knowledge by each new generation of children.

Others have described versions of this amnesia. Pauly (1995), for example, has written of the "shifting baseline syndrome" of fisheries: "Essentially, this syndrome has arisen because each generation of fisheries scientists accepts as a baseline the stock size and species composition that occurred at the beginning of their careers, and uses this to evaluate changes. When the next generation starts its career, the stocks have further declined, but it is the stocks at that time that serve as a new baseline. The result obviously is a gradual shift of the baseline, a gradual accommodation of the creeping disappearance of resource species" (p. 430). Along similar lines, in terms of humans adapting to disease, Dubos (1965/1980) has argued: "Any disease, or any kind of deficiency, that is very widespread in a given social group comes to be considered as the 'normal' state and consequently is accepted as a matter of course within that group" (pp. 250–251).

Solutions to the Problem of Environmental Generational Amnesia

Let us assume, then, that the hypothesis of environmental generational amnesia has merit, and that the phenomenon exists. What can we do to address the problem? I would like to offer six initial suggestions.

1. Get Children Out Into Nature

A basic tenet of constructivist psychological theory is that children construct many deeply held understandings through direct interaction with their physical and social world (DeVries and Zan 1994; Kohlberg and Mayer 1972; Turiel 1983). Many of these understandings become structurally reorganized and increasingly sophisticated in development, as children seek to accommodate discrepant information. But some understandings can more or less stay with us into adulthood, even if they are mistaken. For example, all of us speak about the sunrise because that is what we experience. The sun starts below the horizon, then peeks over it, and then rises high into the sky. Of course, at some point we learn intellectually that it does not actually happen that way. Rather, we learn that the earth rotates. But our direct experience every day—that the sun appears to rise in the sky—continues to confirm our childhood understanding, and I think that is why we prefer to speak of a sunrise rather than an earth-turn. Thus, in an odd sense, environmental generational amnesia occurs through a failure of constructivism. We just plain construct mistaken understandings of what counts as a healthy, vibrant, life-sustaining natural world, and we do not receive enough discrepant information to make it easy to correct our mistaken understandings.

But we can be clever. Because the natural world is not homogeneously degraded, we can use the same constructivist principles in our favor by encouraging children to engage with a natural world that is richer and deeper and less polluted and less degraded than the one they currently experience. That nature can be close at hand. It can be bugs underfoot, piles of leaves, piles of snow, and tall trees. It can exist just outside one's doorway or in a city park. That nature can also be an hour's outing away, or a day away, into a somewhat more pristine area, to help children, especially those in dense urban areas, to calibrate more accurately.

It can work. For example, in a collaborative study I conducted in Lisbon, Portugal, interviewing adolescents and young adults on their environmental views and values (Kahn and Lourenço 2002), one participant told us:

My grandmother lives in the North and I go there. And there are many rivers that still aren't polluted. And I think that, I go up there, and then I come back. I see up there a river that is not polluted. I feel the water running. I come back down here, I see trash, I think that there is such a difference. And I would like that the Rio Tejo—because I live in Lisboa, I was born in Lisboa—would like that the river in my hometown were not so polluted.

Of course, for such experiences to occur we need a more pristine nature for children to experience. Seen in this way, it becomes crucial to preserve more pristine areas in settings both urban and rural. That was one of the ideas I was trying to tell the National Park Service, in the exchange I had with them that I mentioned earlier.

2. Tell Children Stories of How It Was

Interaction with nature usually happens best when it involves both the body and mind; but even the mind by itself can do some work. Consider the experiences of two college-age participants from the Portugal study I just noted (Kahn and Lourenço 2002):

I heard that some time ago, when there was none of that pollution, the river was, according to what I heard, was pretty, there were dolphins and all swimming in it. I think it should have been pretty to see, anyone would like to see it.

I remember, for instance, a person who still talks about the time when he used to swim in the Rio Tejo, and that he misses that a lot. And I, just eighteen years old, find it difficult to believe that this was possible. However, that was the main source of enjoyment of that person.

Granted, such dialogues can fall prey to adult monologues that romanticize the past and gripe about the present and future ("Let me tell you how things were so much better when I was young. . . "). But when such dialogues form part of engaged conversations, they have their place. They provide another means for children to gain information about a healthy and robust natural world. Along similar lines, teachers can use historical diaries and historical novels to convey a sense of the landscape of years past; and writing assignments can involve students in the comparative endeavor: "If you were the person in that historical novel, and you saw the land today, what would you see and what might you say?"

3. Imagine the Future

Imagine we were back in the early 1900s when a few Model T Fords were bumping along on dirt streets, which were otherwise used extensively by horses and people. A Mr. Techne joins us. We crack open a few bottles of soda pop, and talk about the future. Mr. Techne tells us that he thinks he knows what's going to happen within the next hundred years. We ask him to tell us. And he does. He first tells us that our dirt streets will be replaced with brick streets; and we find that a little hard to believe, because it takes a lot of bricks to make a hundred yards of road. But he tells us we're not talking yards, we're talking miles, we're talking entire cities overlaid with a grid of brick roads. Then he explains that a new process will be invented that will produce a product called asphalt. It will be harder than bricks and completely smooth. He then tells us that in less than one hundred years the entire United States will be crisscrossed with paved asphalt roads, going just about everywhere, and that the major cities would be jumbled with roads, and roads over roads, bridge-roads (called overpasses), and overpasses over overpasses, and that at first there would be a wonderful sense of freedom driving on land and over land at speed, but too soon there would be too many cars for even all of these roads, and that there would be something called traffic jams, where sometimes you couldn't drive faster than a man could walk, in cars that pollute the air so bad that some people wheeze. We might take a swig of our pop, and then tell Mr. Techne that that was a good story, but it isn't going to happen that way. We might say, for one thing, paving roads costs an enormous amount of money. How in the world are people going to pay for that? And where are all these people coming from? Your numbers are wacko. And even if somehow all those millions of people magically appear, how can they all afford to buy a car? And, in any case, no one is stupid enough to sit for hours a day in this thing that you call a traffic jam. And finally millions of people would have to be more than stupid to make themselves sick on their air. We might say something like that, but we would have of course been wrong. In the same way, we might have been wrong imagining a future at the inception of other technologies, such as the printing press, television, telephone, computer, and Internet. In the same way, we might have been wrong in not imagining it possible 11,000 years ago that hunter-gatherer life would so suddenly give way to domestic agriculture, or that it would take a mere 1,000 years for people to spread from the then uninhabited land of what is now the U.S.–Canadian border to the tip of Patagonia, or that more Native Americans would die from European germs than on the battlefields, or that China's direct economic losses due to desertification would sit currently at around $40 billion each year (Diamond 1997, 2005).

It is hard to imagine the future. But if we want to solve the problem of environmental generational amnesia, it would be good to do so.

Science fiction writers can help us. I have used them in this book. They are particularly good at playing out different futures, and helping us to envision and feel our lives within them. Through such envisioning, we can better see the impact of the shifting baseline, and change course accordingly.

Or we could forget the fiction, and focus on science, which is perhaps the most obvious of all methods for alerting us to the shifting baseline. I will not belabor this point. But I would like to highlight how the right technological tools, coupled with scientific modeling, can be especially powerful in conveying the shifting baseline on complex problems. My experience has been with a system called UrbanSim, which is a large simulation software system for predicting patterns of urban development for periods of twenty years or more, particularly in terms of how urban development affects land use and transportation (Borning, Friedman, and Kahn 2004; Friedman et al. 2008; Friedman, Kahn, and Borning 2006; Waddell 2002). It has been a decade-long project, principally directed by Paul Waddell and Alan Borning.

Here is one context of use for UrbanSim. Imagine a situation where housing developers want to build a large subdivision in an area ten miles from an existing urban boundary. They explain to the regional planning board that the houses will be LEED certified, that the entire landscaping will follow principles of "biophilic design" (Kellert 2005; Kellert, Heerwagen, and Mador 2008), and that they will even pay to put in the entire two-lane road to accommodate all of the traffic between the subdivision and the urban boundary. The developers say that people will have good places to live. Construction jobs will be created. Tax revenues will go up for the region. Is there a problem? The planning board could do well to ask the question: What happens in twenty years? Here is one possible answer. Once the subdivision is in place, and the road is put in, a small school is chartered, a few grocery stores spring up, and then another subdivision is built nearby, piggybacking on the initial infrastructure of water, electricity, and communication cables. Then three other subdivisions are built halfway between the initial one and the urban boundary, because the road is there. Since it is a two-lane road, traffic becomes a little congested from that halfway point into the city, because almost everyone commutes into the city for higher-paying jobs to pay for their nice homes. More subdivisions then get built, filling in some wetlands. You can see where this scenario goes. It is a micro-version of the Model T scenario by Mr. Techne above. Perhaps strip malls begin to line parts of the two-lane road. Subdivisions fill in along the highway and then branch out on its sides. Traffic becomes snarled. After twelve years, the initial road is widened to accommodate more cars. But then surprisingly the traffic actually becomes worse. It appears that widening the road further increased the rate of expansion into the area. Everyone now plans their travel around the worsening traffic patterns. Also, as cars do, these cars pollute the air, and people start getting used to smoggy days and a little bit more asthma, especially in the children and the elderly. Since the wetlands had been filled in, the birds stopped migrating through these parts. The human-caused erosion and sewage have also diminished the health of the nearby river, and the wild salmon no longer run. That is why people no longer fish it. Recall the college-aged Portuguese student I quoted who said: "I remember, for instance, a person who still talks about the time when he used to swim in the Rio Tejo, and that he misses that a lot. And I, just eighteen years old, find it difficult to believe that this was possible." A version of that now happens in this region, and the young children hear stories from their parents, that twenty years ago when the subdivision had just been built, they used to wake up on a beautiful clear Saturday morning and pull salmon out of the river for dinner that night. Who would have imagined all of these effects when the developers with big smiles and deep pockets proposed their initial subdivision?

UrbanSim is good at helping people imagine future effects. Based on scientifically credible models of how land use, transportation, and human and ecological systems interact, it forecasts plausible scenarios of what a region will look like if certain large-scale decisions are made, such as whether to build a new freeway, expand transit service, or change land use regulations. Central to the UrbanSim software system is that you can query it. You can ask, for example, what would be the effects 5 or 20 years from now if we decided to build an 8-lane bridge instead of a 6-lane bridge over a specific waterway? What if we built a 6-lane bridge but added light rail? What if we kept the current 4-lane bridge as is but increased the bus routes 50 percent and built bike paths? And regarding "the effects"—it is here that UrbanSim seeks to supply data on any variable that people think is important to the decision. Different people will of course bring different interests and queries. A question might sound like the following: Twenty years from now, if we do it the way you propose, will we have more walkable neighborhoods? Will there be space for business expansion? What will be people's commute times from different parts of the region? Will there be affordable housing, and where? Freight mobility is my thing, and I want to know, how easily will we be able to get our products into and out of this region? How much open space will there be? How about the wetlands? Through such queries, UrbanSim is a technological tool that supports the solution I previously discussed (3 above, "Imagine the Future"), that our societal discourse needs to be nuanced, complex, and sometimes contradictory. To date, UrbanSim has been applied in the metropolitan regions around Seattle, Eugene, Honolulu, Salt Lake City, and Houston. There have been research and pilot applications of UrbanSim in Amsterdam, Detroit, Paris, Phoenix, Tel Aviv, and Zurich. UrbanSim also played a significant role in the out-of-court settlement of a lawsuit in Utah regarding a major freeway construction project.

Urban transportation and land use discussions (and resulting decisions) are often emotionally charged. It would be surprising if it was otherwise, for the issues affect people's livelihood and involve their deeply held values. One of the benefits of using UrbanSim is that it helps ground such discussions in scientifically credible outcomes. UrbanSim does not recommend policy decisions, and in principle does not favor any stakeholder over any other. The system has been designed to be largely value neutral. The caveat, however, is that objective knowledge can sometimes lead people more one way than another. For example, in theory, we could all sit around the table, play out the subdivision scenario I just sketched above, agree that such long-term effects are likely, and say "Yep, that looks good to me, I don't mind a little traffic, how

about you? Yep, looks good to me, too. I don't even like to eat fish."
More likely, however, when we look directly into a future with a shifted
environmental baseline, it is not so easy to accept.

4. Be Cautious in Believing That New Technologies Can Save Us

In chapter 3 I suggested that there appears to be a prevailing worldview
that is protechnology. This view holds that if an interesting technology
can be conceived of and built, go for it; and if later there are problems,
we can figure out a technological solution then. I also agreed with
Diamond (2005) that new "technologies, whether or not they succeed in
solving the problem that they were designed to solve, regularly create
unanticipated new problems" (p. 505). Also in chapter 3 I discussed how
the mechanization of cigar making in the early 1900s caused unanticipated problems for the life of the factory workers and undermined the
tradition of the lector. I also discussed the idea of something gained,
something lost, with such technologies as the printing press, electric
lights, and airplanes. In this chapter, I have discussed some of the unanticipated problems of cars. There are thousands of other examples. This
mistaken worldview can wrongly dismiss the problem of environmental
generational amnesia by saying that although things look like they will
be bad in the future, new technologies will save us; so "don't worry, be
happy." That reasoning did not work for technological societies in the
past that failed because of their poor environmental practices. There is
no compelling reason to think that we are in a privileged position. True,
our technologies have become increasingly sophisticated and powerful;
but then so too have the unanticipated problems they create. Think of
nuclear power. It helps to solve our need for energy. But a Chernobyl
meltdown or worse is not a pleasant event, and storing spent radioactive
fuel safely for thousands of years may always require a large measure of
luck. Again, I am not advocating the Luddite position. But, in this
context, I am saying that we should not believe that future technologies
can make light of the problem of environmental generational amnesia.

5. Conduct Research on the Effects of Nature and Technological Nature on People's Physical and Psychological Well-Being

This point will be my briefest of this group, because it is already the
longest. It comprises the seven empirical chapters of this book.

12

Adaptation and the Future of Human Life

Through creating and using technological nature, we are changing long-standing patterns of interaction that have existed for tens of thousands of years. Is that a problem? Some people say "No, it's not a big deal." At the core of their argument usually lie three claims, which sound something like this: (i) "We'll adapt." (ii) "Adaptation is how we evolved." And (iii) "Adaptation is good for us." The first claim is true. We will adapt. It is that or we will go extinct, and I doubt that will happen. The second claim is also true. Adaptation is part of our evolutionary heritage. But the third claim is not always true. It is possible to adapt and then to diminish the quality of human life, if not suffer. Let us imagine, for example, that all of us were put in prison for the rest of our lives. If we did not die straight out, and most of us would not, we would adapt to our new environment. We might get fatter from lack of exercise. We might become more violent. We might shut down emotionally as a way to cope, and hardly be aware of it.

How will we adapt to technological nature? How can we make wise choices in directing its creation and how we use it, now and in future years?

So far in this book, there have been three parts to my answer. First, I have argued (chapters 1–2) that we need rich interactions with nature for our physical and psychological well-being, but that we are losing those interactions because we are quickly and pervasively degrading if not destroying large portions of nature, which are required for such interaction. Second, I have presented and discussed my collaborative research on the psychological effects of experiencing technological nature (chapters 4–10). Taken together, this body of research suggests that technological nature is better than no nature but not as good as actual nature. Third, in my account of environmental generational amnesia (chapter 11), I have suggested that through the loss of experience with actual nature, and the increase in technological nature, the baseline for what counts as a healthy environment, and healthy psychological functioning in that environment, will shift downward, as it has already, as each generation constructs a new, more impoverished baseline of what is normal. Environmental generational amnesia is a form of adaptation. Each generation adapts psychologically to a new environmental baseline.

Humans adapt to their environments, however, in many different ways. The adaptations can involve different processes and mechanisms, occur at the level of an individual, culture, and species, and have consequences that last for seconds to many thousands of years. Thus, to understand how we will adapt to technological nature, we first need to venture into difficult conceptual territory that involves what adaptation is and how it works.

Adaptation Is Not by Itself Normative

Many people tend to believe that adaptations are good and lead to better biological systems. They likely believe so because that is how adaptations often appear to work. For example, when we enter a darkened area, the pupils in our eyes adapt by dilating; when we then reenter a brightened area, the pupils adapt again by constricting. In the summer, people with fair skin tan, which is the body's way of adapting and thereby increasing its resistance to harmful ultraviolet radiation. Muscles adapt to strenuous efforts and get stronger. When we go to higher altitudes, our bodies adapt in terms of changes to our breathing, blood hemoglobin, and tissue metabolism. In hot conditions our bodies sweat to regulate their core temperature. Our bodies adapt when we change time zones. Entire cultures adapt to their climates. In the arctic, for example, we see more of a stocky and fat body type, which has been adaptive, as that body type helps to conserve heat.

Some of these examples draw on a biological evolutionary conceptualization of adaptation: Genes that lead to body types or behaviors that enhance survival tend to reproduce themselves, and thus these genes and correlative behaviors grow more frequent. For example, in the above example of arctic cold, it would be hypothesized that people born with a stockier and fatter body type had a slighter better chance of survival in that climate, and thus over time, that body type tended to prevail. In chapter 2, I discussed how Wilson (1984) uses this biological adaptive explanation to argue for the existence of biophilia. Namely, certain affinities for nature (such as for water and plants) and fears of nature (such as of snakes and heights) increased people's chances of survival, and thus those affinities became more prevalent in our species.

It is true that biological adaptations can be good for us, as humans. But it is essential for us to understand that such adaptations can also be bad for us, or simply neutral. I would like to develop this idea.

First, we should set aside any thought that adaptations lead to *moral* goodness. To think so would be to commit what in moral philosophy goes by the term the *naturalistic fallacy*. Its specification is often credited to Hume (1751/1983) and Moore (1903/1978). The fallacy can be stated simply. "Is" does not equal "ought." For example, let us for a moment agree with Freud (1930) that "aggression is an original, self-subsisting instinctual disposition in man" (p. 77). That biological quality by itself would not establish the moral justification for its expression, such as for killing one's neighbor to gain his property. Another example is that in the 1930s Nazi Germany appeared to be adapting quite well to the world environment; but such success in their adaptation did not make that adaptive behavior morally right. Similarly, when Europeans came to the United States, they wiped out most of the Native American populations and took their land, and in the centuries since have successfully adapted to the new land; but that adaptation does not by itself establish the moral legitimacy of the invasion and the killing of many native people. In short, if you want moral justification you have to do something more than simply establish the existence of a trait or behavior that has been adaptive.

Morality aside, adaptations can be not only good but bad for the biological system. In the previous chapter, I discussed a study by Evans, Jacobs, and Frager (1982) that found that people who live with a certain level of air pollution for an extended period of time become desensitized to that pollution, and less readily recognize that such pollution exists. Imagine, then, that we had grown up in a rural area with clean air and then as adults moved to a dense, noisy, and polluted city, such as Los

Angeles. Over time, we would become desensitized to, and in this sense adapt to, our new polluted environment; but that environment would still be harming our health. Imagine, now, that we then took up residence close to a noisy freeway in Los Angeles. Perhaps initially we would have trouble falling asleep at night. The noise would be too loud. But over time we would likely get used to it, and in this sense adapt to it, and hardly notice the noise. But that does not mean that the noise would still not act as a stressor on our biological systems. I noted earlier that if we were forced into a prison for the rest of our lives, we would adapt to that new environment, but we would not enjoy it, and we would fare worse than we do now—much worse. Another example is that some people eat fast food corporate style: French fries, Big Macs, Whoppers, chicken nuggets, bacon-burgers, sodas, milkshakes, fried chicken. They like the taste of this food. They need it. Their bodies have adapted to this food. If they do not eat it regularly, their bodies call out for compliance. But as we all know, fast food is not good for human health.

I am saying that one misconception people sometimes have is that adaptation to the environment is always good for the biological system. It is not. A second misconception is that although an adaptation might be harmful for any particular individual in a society, adaptations are always good for the society at large. But that also is not true. For example, we can get used to bad traffic in our urban environments, but such traffic can still be stressful for millions of people, waste time, cost money, and contribute to air pollution and global warming. In other words, that traffic that we are getting used to is not good for society.

Even when adaptations seem good for a society, the question needs to be asked: "Over what time frame?" It can happen, and often has, that the adaptive behaviors that allow a culture to thrive for 100 or 1,000 years are the very behaviors that cause their demise. For example, as I described in the previous chapter, humans arrived on Easter Island by 900 CE, and they adapted quite readily to the subtropical forest they encountered. As part of their adaptive behavior, they extracted natural resources at an unsustainable rate. The upshot was that although that behavior was highly beneficial to the individuals and people of Easter Island over some hundreds of years, because their behavior was not sustainable, it led first to hard times and then to horrible times, during which many people practiced cannibalism to survive, and finally it led to the end of them as a people. Thus what might look like a good adaptation from a shorter time frame can look bad from a longer one.

Adaptation: What Is It?

How then do we know if an adaptation is good, bad, or neutral? Part of my answer above is that the answer can vary depending on the time frame under consideration. I will come back to this idea in the next section. Another issue I will address later in this chapter is what exactly counts as a "harm." But before addressing both ideas—how long harms last and what counts as a harm—first we need to tackle a more fundamental question: How exactly do we define the term "adaptation"? Unfortunately, there is no easy answer. As Siegler (1996) writes: "Adaptation is perhaps the single central organizing idea in modern evolutionary theory. . . . This is true despite the concept being almost as difficult to define in biological as in psychological contexts" (p. 26).

Merriam-Webster's dictionary ("Adaptation," n.d.) provides a worthwhile starting point. It defines adaptation in two ways. One way simply refers to an organism's "adjustment to environmental conditions." This definition says nothing about whether the adjustment benefits or harms or neutrally affects the organism's system. The second way casts the adjustment positively, as "modification of an organism or its parts that makes it more fit for existence under the conditions of its environment." But as I sought to establish in the previous section, adaptations do not always lead to greater fitness. Thus, in my view, Webster's first definition has the most merit. Adaptation refers to an organism's adjustment to its environment and makes no normative claim.

We still, however, have not said too much insofar as there are many ways that a system adjusts (adapts) to environmental conditions. Homeostasis is one form of adaptation wherein an organism seeks a stable "normal" state, while being able to handle some minor variation. Our bodies, for example, seek to regulate their internal temperature to one that is normal (e.g., 97.8 degrees Fahrenheit). Sometimes a body is hotter, especially when it is fighting a disease with a fever. Other times the body's temperature might drop a little lower. But sooner usually than later the temperature comes back to normal. Something like this process occurs when we travel to different time zones and our bodies adjust our internal clocks to match the sun's circadian cycle. One difference here, however, is that it is also possible to decide not to adjust to the circadian cycle, as do people who work night shifts. But even then the body usually performs better sleeping at night and being awake during the day. Some form of homeostasis is involved in acclimatization, when a body seeks to normalize its breathing, blood hemoglobin, and tissue metabolism when it changes altitudes. If the altitude increases too much, the adjustments are not fully successful insofar as a new and impoverished baseline gets established. Although people can live their lives with that new baseline, they are poorer for it—in terms, for example, of how fast wounds heal and the capacity for physical exertion. Over many generations a different form of adaptation can occur, and a people can change genetically to function "normally" at higher altitudes, as have the Sherpas in Nepal. But there are obvious limits. No one could presumably adapt to living their lives at elevations equivalent to the top of Mt. Everest.

Addiction is also a form of an adaptation. Webster (1973) defines addiction as "a compulsive physiological need for a habit-forming drug (as heroin)." But what counts as a drug? We would all agree about heroin. How about tobacco? Caffeine? Are people addicted to Ibuprofen if they use it daily to manage low-level pain in their body? Does it make sense to say that some people are addicted to chocolate? To Big Macs? To soda? All of us have a compulsive physiological need to eat food. Are we addicted to food? People have also used the term "addiction" to speak about compulsive needs not just to drugs or ingestible items but to forms of activity. People speak, for example, of an addiction to gambling, especially when there appears to be something compulsive to the behavior. But how far are we willing to go with this move? Do we also want to speak of an addiction to shopping? An addiction to following one's football team? An addiction to falling in love? With all of these questions, notice what happened. We started with a term that specified the parameters of a form of adaptation: addiction. The specification initially seemed straightforward. We all agree that people can get addicted to heroin. But then the boundaries of the definition quickly got blurred.

The same thing happens with the term "habituation." Webster offers two definitions. The first overlaps substantially with addiction: "tolerance to the effects of a drug acquired through continued use." Thus, as they did with addiction, blurred boundaries emerge with this definition. Would we say, for example, that we become habituated not just to chocolate or fast food, but food in general? If so, we are not saying very much. If we want more specificity, then we need once again to demarcate what counts as a drug, which is not easy to do. The second definition is very broad: "the act or process of making habitual or accustomed." Drugs aside, clearly we adapt to our environment all the time in this way. We become accustomed (habituated) to sitting through Sunday morning church or long Passover Seders, to army life, to bad traffic, to an abusive relationship, to sunny days if we live in Phoenix and rainy days if we live in Seattle, and to the fast flow of information over the Internet. Taken this far, we become habituated to virtually anything and everything within a normal range in our experience. Psychologically, it is also possible to become habituated or acclimated to difficult events outside a normal range of experience. For example, think of an individual who progresses through a life-extracting disease, such as AIDS. Over months and often a handful of years, the individual succumbs to increasingly awful physical states, and at each of those degraded physical states, the mind needs to adapt in the sense of readjust.

Adaptation takes on a different meaning when applied to a child's cognitive development. According to Piaget (1983), a child brings to new situations an existing way of understanding them, and in this sense seeks to assimilate the new to the old. But that process never works completely, so the child also needs to accommodate to the new. Sometimes the accommodations are not successful. At that junction, the child is disequilibrated, recognizing the problem but not the solution. Thus, according to Piaget, toward seeking equilibration, and through interaction, the child reorganizes existing structures of knowledge to take account of the new and previously discrepant information. Notice, then, that in this account of the equilibration of cognitive structures the psychological system does not seek homeostasis in terms of an original state (as occurs with the body's homeostasis around its normal internal temperature) but in terms of new and more comprehensive and adequate psychological structures.

This account of disequilibration and the construction of knowledge grounds the speculative thoughts I offered in chapter 8. Recall that I had said that children in the coming decades will be interacting with a new form of an artifact that has never existed before in human history—one that is technological but seemingly autonomous, goal-directed, personable, and social, and perhaps even moral: personified robots. Since children's conceptual categories emerge not simply by maturation but through interaction with the environment, I then speculated that children will construct new ontological categories of knowledge. In other words, in the classic Piagetian telling of equilibration, the child's developing mind constructs categories that are already fully established as core knowledge that adults have. But in this new telling of the equilibration, the child's developing mind—as it adapts to the new technologies—will construct new categories of knowledge that all of us who are old enough to be reading this book do not and will never have.

As I have described these different forms of adaptation, it might appear as if they always occur in isolation from one another. But often they are combined. For example, the adaptation that is involved with environmental generational amnesia has aspects aligned with equilibration, habituation, and homeostasis. It is aligned with equilibration insofar as it depends fundamentally on the child's construction of knowledge through interaction with the physical world. It is aligned with habituation and homeostasis insofar as the psychological system seems strongly committed to the "normalcy" of the baseline that is constructed in childhood.

In summary, I have suggested that it makes the most sense to define adaptation in terms of an organism's adjustment to its environment, and that such adjustments can help or hurt a biological system, or both help and hurt, or neither. In other words, adaptation is not in itself normative. Then, without attempting to be comprehensive, I showed that adaptations occur through homeostasis, acclimatization, addiction, habituation, equilibration, assimilation, and accommodation. These different constructs provide rich accounts of different ways that humans adjust to their environments; yet the constructs themselves often seem to share overlapping and underspecified processes and mechanisms, and to comprise poorly defined criteria for bounding the application of the construct.

What's So Bad about Being a Dog Anyway? or, How Long Do Harmful Adaptations Last?

Earlier I said that what might look like a good adaptation from a shorter time frame can look bad from a longer one. Thus, our next step—toward answering the question of how we will adapt to technological nature—is to try to bracket what some of these time frames look like. Here, then, are five time frames that I find useful for the discussion at hand. I should say at the outset that I have little vested in the number of levels or the spans of time within each. If the reader prefers to rethink this broad typology with different levels and time frames, that is fine, as my point is more to lay out the structure of the argument than to commit to the specifics of it. I also remind the reader that I have not yet specified what counts as a harm, and I am for now emphasizing physical harms (such as pain, sickness, and death) because of their prima facie validity.

Level 1: Harms Last from Milliseconds to No Longer Than About One Month

Many of the adaptations noted above have harms of this duration. For example, when we move from a darkened room to bright sunlight, it hurts our eyes, and our pupils constrict. The harm is in terms of our initial pain. The pain lasts on the order of seconds. When we fly halfway around the world, we experience the harm of jet lag. Within a few days or a week our bodies adapt and the harms disappear. Similarly, when we proceed quickly from lower to higher elevations, from, say, sea level to 10,000 feet, most people experience the harms of altitude sickness. Within a day or at most a week, our bodies adapt and the harms disappear. Notice what happens when we ask the question: Is it harmful for the body to adapt as it moves quickly from a darkened environment to one that is bright sunlight? The answer is yes and no. It is harmful (as in a little painful) over the duration of some seconds, and it is not harmful past that. Similarly, we can ask the question: Is it harmful for the body to adapt as it moves quickly from sea level to 10,000 feet? Again, the answer is yes and no. It is harmful (at least to most bodies) over the duration of a day, and it is not harmful past a week.

Level 2: Harms Last between One Month and a Lifetime, but Future Generations Experience No Harms

When wild animals are brought into captivity (as in a zoo), they often exhibit "neurotic" behavior: for example, a large cat can pace for hours each day in its cage, and an elephant can stamp its foot for hours each day in its cage. These are harms to biological entities that can last a lifetime. Often because of these harms it is difficult to get wild animals to breed in captivity. But when breeding is successful, and the next generation is born into captivity, that next generation often adapts much more easily to their captive environment. The same can occur with humans. It can be difficult for people who grew up rurally to adapt as adults to a dense crowded urban environment. In some ways, they may be like the wild animal in the zoo. But their children will likely have few of the same difficulties as their parents. Using our earlier terms, the parents are unable to habituate to their new urban environment, whereas their children, born into that environment, habituate at a young age. We can ask the question: Is there a harm when people move from rural to urban environments? If for the moment we assume the harms are of the form I noted above, then the answer is yes and no. It is harmful to the parents for the entire duration of their lives. But it is not harmful to their progeny across future generations.

Level 3: Harms Last between 1 and 10 Generations

Plagues offer one of the clearest examples of harms at this level. Throughout history, epidemics have wiped out large percentages of a people within a few years, or even a week. But then over a period of years—without any medical vaccines—resistance to the disease builds up within a society, and the disease either substantially diminishes or disappears altogether. For example, between 1485 and 1551 the highly contagious English sweating disease struck England and parts of Europe, and then "it vanished as swiftly as it had come" (Dubos 1965/1980, p. 187). During roughly the same time frame, syphilis spread through Europe at an astonishing rate, and then decreased (Dubos 1965/1980). One of the deadliest pandemics in human history was the Black Death, which killed around 75–100 million people. In the fourteenth century, it is estimated that in Mediterranean Europe, Italy, the South of France, and Spain that the Black Death killed 75–80 percent of the population. In Germany and

England, the number was probably closer to 20 percent; in Egypt, about 40 percent; in Iraq, Iran, and Syria, about 33 percent ("Black Death," n.d.). Most of the damage was done within about four years, at which point the plague largely retreated, although it is believed to have returned more locally every generation for several hundred years. In 1665 in London, for example, the Black Death erupted and killed at least 100,000 people. By the 1700s, however, even such local outbursts subsided. For all of these plagues, there is a common adaptation account. People who were particularly susceptible to the disease got it, and they either died or lived and built up immunities to it. Most died. The children of those people who lived thereby had a genetic predisposition to be resistant to the disease. It is a classic case of biological evolutionary adaptation.

The reason I chose 10 generations for the outside mark of a Level 3 harm is that if we think of the Black Death as the worst plague in our recorded history, and if it lasted roughly 200–250 years (although only about 4 years at its worst), and if each generation spans roughly 20–25 years, then 10 generations roughly accounts for the length of this plague. After 10 generations, this plague caused us as a species no further harms. None at all. It was not due to medical science coming up with a vaccine or a cure. It was due to biological adaptation.

Now we ask the question: Are plagues harmful? Once again the answer is yes and no. They are tremendously harmful over the short and medium term (Levels 1–3). They kill many people. But they are not harmful to us as a species over a longer time frame.

Level 4: Harms Last between 10 and 500 Generations

Humans are susceptible to a huge number of diseases. For example, we can be struck with ulcers, diabetes, allergies, chicken pox, hives, lupus, syphilis, psoriasis, ringworm, basal cell carcinoma, AIDS, migraine headaches, cholera, genital warts, gonorrhea, shingles, gout, herpes, measles, scarlet fever, smallpox, thrush, typhoid fever, and Alzheimer's disease. We can also be born with or come to have many physical disabilities. We can have poor eyesight, poor hearing, a cleft lip, or a weak heart. We could end up with cerebral palsy or multiple sclerosis or muscular dystrophy or Parkinson's disease. We could be born with Down syndrome. For some of these diseases and disabilities, we now have medical cures and technological solutions. For example, for most people with poor eyesight, the solution is very simple. They wear glasses and they see normally. People with poor hearing can wear hearing aids. Surgery can partly correct for people with weak hearts. Insulin can help people with diabetes. We have reasonably effective vaccines for many diseases such as chicken pox, hepatitis, tetanus, mumps, measles, tuberculosis, and typhoid. And so on.

The short of it is that without medical and technological assistance, many of us would not have been born because our ancestors had defects that would have killed them before their childbearing years, or we would have died before now due to our own defects. As Dubos (1965/1980) writes: "It is to be expected that medical advances will modify the hereditary structure of prosperous populations by permitting the survival of persons with genetic defects who in the past had little chance of reproducing their kind" (p. 248). Thus, as we use our technological minds to help us and others as individuals, we as a species diminish our natural capabilities and become increasingly reliant on technology to survive. In this way, over the long term—roughly 10–500 generations—as we care for specific individuals we harm ourselves as a species.

Within this longer time frame, and because of the construction across generations of a shifting baseline, we as a species will unfortunately not recognize many of these harms as harms. This mechanism underlies my account of environmental generation amnesia in the previous chapter. Instead, we will largely think of these conditions as the normal human condition. Even today we do so. For example, I doubt any of us is ever surprised or particularly concerned to see others or ourselves wearing eyeglasses. As Dubos (1965/1980) writes: "Any disease, or any kind of deficiency, that is very widespread in a given social group comes to be considered as the 'normal' state and consequently is accepted as a matter of course within that group" (pp. 250–251).

Level 5: Harms Last Over 500 Generations—or Disappear

The average size of a gray wolf's territory is about 77 square miles. The wolves, as a pack, cover about 15–30 miles a day (about 9% of their territory) in search of prey. When chasing prey, a gray wolf can reach speeds approaching 40 mph, bounding as much as 23 feet at a time. The wolves communicate the boundaries of their territory to other packs somewhat by scent marking and principally by howling ("Gray Wolf," n.d.). Now imagine a gray wolf held captive in a zoo, pacing from one end of its cage to the other repetitively for hours on end. It is hard to escape the interpretation that the wolf in a zoo is living an impoverished life.

Over the last 15,000 years, a form of the wolf has been domesticated. This form of wolf has changed genetically. We no longer call this domesticated wolf a wolf. It is now "man's best friend" and goes by the name of "dog." Does the dog suffer by not being a wolf? No, because the benchmarks have changed. Do we suffer because we are no longer single-cell prokaryotes or *Homo habilis*? No, because an organism's well-being needs to be evaluated based on its capacity for flourishing for the species it is, not the species it once was. Now we ask: If over 500 generations (or pick whatever time frame you think it would take) we adapt to impoverished environmental conditions in such a way that we change as a species, in terms of our genotype and phenotype, will we be harmed? Following the logic of the argument, the answer would be no.

Notice, then, the difference between a Level 4 and Level 5 harm. In a Level 4 harm, we are harmed but do not necessarily know it because of the shifting baseline and resulting generational amnesia. The benchmarks are tied to the organism's physical and psychological capacity, and that capacity is independent of whether or not the organism correctly knows what that capacity is. In a Level 5 harm, the organism can fundamentally change and become another species, at which point the benchmarks change and the harms disappear.

Imagine that we enslaved all the people of the world who had big noses, or white skin, or black skin. We can call this group People S (for Slaves). All People S would incur Level 1 harms. In the initial process of being enslaved, some People S would rebel, and they would be killed. Thus, those rebellious traits would start to be selected out of the population. Other People S would have children. Those children would be born into slavery, and thus would have an easier time than their parents adjusting to their enslaved condition. But still those children and their children and many future generations to follow would suffer in slavery. These People S would still yearn for freedom, autonomy, and self-governance, as enslaved people throughout the world always have. The story of Exodus is a story of an enslaved people yearning for and gaining such freedom. But let us say we could keep the People S enslaved

simple but lovely form of interaction with nature, which often begins in childhood, that involves collecting small objects from places that one visits. Sometimes children build a large collection of such objects, and classify them, and study them. Such forms of interaction can set into motion a lifelong scientific inquiry into the natural world. Sometimes these objects, for children and adults, hold important memories of special times. The new "environmental" message—"take only memories, leave only footprints"—helps to prevent harm to an ecosystem, but it comes at a human cost, not large, but not so small either, by causing a harm of unfulfilled flourishing: the experience and satisfaction of collecting parts of nature.

A different harm of unfulfilled flourishing can be found in Diamond's (2005) account of how the Japanese in the 1700s solved one of their environmental problems: the overharvesting of their timber. Employing what Diamond calls a top-down management style, the local rulers, both the shogun and the daimyo, dictated who could do what in the forests, and where, and when, and for what price. Toward making thoughtful decisions, the rulers paid for detailed inventories of their forests. Diamond writes:

Just as one example of the managers' obsessiveness, an inventory of a forest near Karuizawa 80 miles northwest of Edo in 1773 recorded that the forest measured 2.986 square miles in area and contained 4,114 trees, of which 573 were crooked or knotty and 3,541 were good. Of those 4,114 trees, 78 were big conifer (66 of them good) with trunks 24–36 feet long and 6–7 feet in circumference, 292 were medium-sized conifers (253 of them good) 4–5 feet in circumference, 255 good small conifers 6–18 feet long and 1–3 feet in circumference to be harvested in the year 1778, and 1,474 small conifers (1,344 of them good) to harvest in later years. (p. 301)

Diamond believes this form of management is exemplary. Granted, it was effective in preventing direct environmental harms caused by over-harvesting timber resources. But now we can ask: Do harms of unfulfilled flourishing arise through interacting with such managed land where literally every tree has been counted, measured, graded, and fit into a harvest plan for eventual cutting? I think that such harms do arise. Dean (1997) writes that "an enveloping wild landscape . . . [is] central to our original understanding of the world and our rightful place within it" (p. 17). I developed this idea in chapter 1 in relation to the Ju/wasi of the Kalahari Desert. I wrote that wildness for the Ju/wasi did not just exist facing off a lion with a burning branch. It was encountered—and still can exist for us today—while experiencing the migration of birds, the changing of the seasons, and heat and cold. Wildness is the freedom to move, and the strength to do so, and the land to do it in. Wildness is encountered through interacting with states that are vast, free, and self-organizing. If Dean's position is correct, and I argued in chapter 1 that it is, then perhaps by interacting with heavily managed landscapes we do not experience a sense of awe in the Other—that which exists outside of human domination. We do not experience a sense of humility. Perhaps it is reasonable to say that when we look at the Other and see only a reflection of ourselves, we have not fully experienced ourselves. We have not fully recognized our potential as individuals or as a species.

In chapter 4, I described how the plasma display window of nature came up short, in comparison to its glass window counterpart, in terms of its physiological and psychological benefits. I said then that the technologist can always provide a rebuttal of the form: "But the technology is not yet quite good enough." I said that that was a fair rebuttal, as far as it went, but that as technologists develop new technologies we need to put the technologies "to the test," and that the important question becomes, what tests? I then concluded that conceptualizing the right set of psychological benchmarks would lead to the right tests and constituted a critical endeavor for engineering technological nature for human good.

We are now in position to say more about the right sort of psychological benchmarks insofar as they comprise the above two types of harms: direct harms and harms of unfulfilled flourishing. Most of the harms I have focused on in my empirical research program on the psychological effects of interacting with technological nature have been direct harms, because these are the most tractable scientifically. The most notable exception occurred when my colleagues and I spent substantial efforts trying to establish some robust measures of creativity (chapter 4). Although we were not entirely successful, I thought at the time, and I still believe, that that endeavor represents an important one because it can function as a placeholder for an empirically tractable harm of unfulfilled human flourishing. But what if instead of just the benchmark of creativity, or the few others described above—such as collecting small parts of nature or interacting with unmanaged landscapes—we could successfully map out hundreds of benchmarks for human flourishing in human–nature interaction? If that were possible, we would then have much of what we would need in terms of criteria by which to assess the psychological effects of technological nature in its current and future instantiations. That is the agenda I put forward in the next section.

A Nature Language—Benchmarks for Human Flourishing

Imagine the following situation. You live in a town with access to a beautiful river, and have on many occasions meandered down river off-trail over sharp rocks and small waterfalls to secluded spots of immense beauty. There is one pool in particular that takes effort and time to reach. You do not always have the time and energy to head there, but when you do, you feel yourself, with each step, pulling away from the busyness of the day and the comfort and safety of the town, and moving toward something less trammeled and more wild. You feel joy and also a little fear. It would not be a good time, for example, to twist an ankle. Hours later, on your return, you look forward to rejoining your family and friends in the comforts of your home in a domestic landscape. Now imagine—which is easy for me to do, because something similar has happened in my life—that people put in a quick access trail directly down to that river pool, so that within minutes anyone from town can enjoy the pleasures of swimming in that special spot. People make a strong case that their lives are busy and there is not always time to walk the long way down river. People also argue that it is elitist to restrict the river pool to those who have the time or the stamina to reach it. Perhaps they even mention that just yesterday they took some young children to that pool, using the quick trail, and that it was a joy to see the children connect to nature in such a way, and who would want to deny children such an intimate connection with such a beautiful nature spot in the world?

What can one say in response? One answer, of many, is that something deep and profound occurs in the human psyche as it moves out and away from human settlements. Often the mind quiets itself from social chatter; the senses become more alert because one is off the trail, finding one's own way, and because you know you need to keep yourself safe. It is not that one is antisocial. Not at all. It is that part of being deeply social is to separate at times from the larger society, and then, from that stance of separateness, to rejoin society. Milton (1674/1978) writes in *Paradise Lost*: "For solitude sometimes is best society, And short retirement urges sweet return" (Bk. IX, ll. 249–250). It has been this way for *Homo sapiens* for tens of thousands of years as part of the life of the hunter-gatherer. This form of interaction with nature has also been, in some form, incorporated into virtually all cultures, as in the adolescent initiation rites among indigenous groups, such as the Dagara of Burkina Faso (Somé 1995). We can refer to this form of interaction with nature as *movement away from human settlement—and the return.* This movement can happen in small groups, as happened in ancestral times when small hunting parties would separate from the main group for one to five days at a time. This movement occurs perhaps most powerfully alone. When Jesus spent forty days and forty nights in the desert, he was not with a support group of other humans.

It was asked above by the advocates of the new river trail: Who would want to deny children intimate connection with beautiful spots in nature? But I think that question is framed incorrectly. First off, the question is part of a slippery slope. Why stop with a trail? Why not put in a driving road, with bus access, and make it wheelchair accessible? Why not put in roads to the top of Half Dome in Yosemite and to the base of Annapurna? Why not replace all trails in wilderness areas with roads, so as to make those beautiful spots accessible to all? Many people would object to such roads but do not see that after 50,000 years of expanding across this planet, we are closer than not to that road-full condition. But even setting aside the counterargument of the slippery slope, the trail advocate's question is framed incorrectly because it assumes that by granting easy access to places in nature we lose nothing in the process. But easy access deprives people of this opportunity to experience the movement away from human settlement—and the return. That trail causes a harm to people now and to future generations. It is a harm of unfulfilled flourishing. People may not recognize this harm. If that is the case, then it likely represents more evidence of the shifting baseline and the problem of environmental generational amnesia.

As we populate the planet with over 6 billion people and increasingly use our technological prowess to control if not destroy nature, or to mediate it, as with technological nature, we are losing patterns of interaction with nature that have sustained us for tens of thousands of years, and which contribute deeply to our flourishing as individuals and as a species. As we lose these patterns of interaction, we are losing the very conceptualization and language to speak about that which we are losing.

Thus there is the need to generate what I am calling a *nature language*: a way of speaking about patterns of interactions between humans and nature, and their wide range of instantiations, and the meaningful, deep, and often joyful feelings that they engender. I believe the vocabulary of this nature language is partly comprised of hundreds of interaction patterns. The above vignette motivated one such interaction pattern: movement away from human settlement—and the return. One of my examples in the previous section comprises another interaction pattern: of collecting objects of nature. I structured chapter 1 around six broadly framed interaction patterns of the Ju/wasi of the Kalahari Desert. Namely, the Ju/wasi used their bodies vigorously. They experienced periodicity of the natural world and in the satisfaction of their physical needs. They had freedom of movement. Some of their desires were checked and balanced by the environment. They encountered the wild. And they cohabited with nature. I said at the end of that chapter that we should care about these interaction patterns because they point not just to our past but—in integration with our technological selves—to our future.

My colleagues and I are currently seeking to generate this nature language, based partly on conceptualizing and validating what may be several hundred interaction patterns (Kahn et al. 2010; Kahn et al. forthcoming). Other examples of possible interaction patterns include the following: traveling the winding path, traveling off path, the hunt, waiting, prospect, refuge, investigating, artistic expression, solitude, approaching carefully, in the flow of nature's dynamics, water on feet and hands, immersed in water, plunging into water, moved by water, playing, dying, gardening, foraging, tracking, combating the destructive forces of nature, using nature to find respite from nature, climbing, running, following the light through a thicket, around a campfire, and under the night sky. Each one of these possible patterns has stories attached to it, in the way that the pattern above (moving away from human settlement—and the return) had a story attached with it, that fleshed it out and gave it a meaning in context.

It is a hard project, and it may be another five years before we can say with conviction what this language sounds like in a comprehensive way. But it would be worthwhile here to describe a few more properties of this nature language, and what this project can achieve, if successful.

Our account of interaction patterns draws in part on Christopher Alexander's pattern work in architecture wherein, over decades, he and his colleagues generated 253 patterns in the built environment that they believe engender meaningful human living (Alexander et al. 1977). For example, one of their patterns is titled "light on two sides of every room." They write: "The importance of this pattern lies partly in the social atmosphere it creates in the room" (p. 748). They also write that this "pattern, perhaps more than any other single pattern, determines the success or failure of a room" because "when they have a choice, people will always gravitate to those rooms which have light on two sides, and leave the rooms which are lit only from one side unused and empty" (p. 747). For Alexander, and us, patterns are not rigid molds, as in those made by cookie-cutters where every cookie looks identical. Rather, patterns embody an underlying unified structure that allows for infinite instantiations. It is in this sense that Alexander (1979) writes that patterns always have a "quality without a name," which "let our inner forces loose, and set us free" (p. x).

So, too, with interaction patterns between humans and nature. The above pattern of moving away from human settlement—and the return does not specify where or how or how long or with whom. Each time it is different, unique, alive. Or consider a possible interaction pattern of *using nature to find respite from nature*. Thomas (2006) has a wonderful picture taken about 1950 of her family sitting with Bushmen in the Kalahari Desert. They are all sitting in the sand in a squiggly sort of circle. Thomas points out that the odd shape of the human configuration follows perfectly the shade from the tree above them. Similarly, this interaction pattern emerges in modern society when we step under a tree to find a brief respite from a thunderstorm or a hot summer sun; or when children sit in a groove among jagged forms of a rock in New York City's Central Park. Or consider a well-known and well-utilized interaction pattern from landscape architecture: the winding and countered path (Kaplan, Kaplan, and Ryan 1998). Such paths lead one forward in anticipation of what comes next, and engender human interest and engagement. This interaction pattern can be instantiated in endlessly different ways. The paths may move like animal trails throughout a forest, or be constructed in an urban park or in a Zen garden. Rivers can be paths in this way as well. Walk up a winding river, and it is likely that your mind will say "oh just go one more turn around the bend, just to see what's next." Then you do. Then your mind says the same thing again: The interaction pattern is at work.

Notice another structural feature of an interaction pattern. The pattern is characterized not simply in terms of the physical features of the natural phenomenon (such as a winding river) nor simply in terms of the human activity (such as walking) but the specific interplay between both. Interaction implies human activity, often physical and always mental with external phenomenon. In turn, the interactions lead to meaningful if not profound emotions, such as of pleasure, contentment, joy, fear, companionship, surprise, delight, reverence, awe, humility. Sleeping under the night sky can fill one with awe, and walking in grizzly country can fill one with humility, fear, and reverence, in ways that no urban experience can.

Another fundamental idea about interaction patterns is that many of them can be instantiated in three categorically different ways: in a wild form, a domestic form, and a perverse form. As a case in point, reconsider the interaction pattern I described at the beginning of chapter 1: being recognized by a nonhuman other. It is a powerful and meaningful experience—one that people sometimes remember for a lifetime—of being recognized by an animal in the wild: perhaps getting within feet of a shark or a barracuda while snorkeling in ocean waters, or of surprising a mama bear and her two cubs while hiking in the forest, or even encountering a large buck in a meadow glen, and there is an instant of silent stillness where he looks at you and you look at him, and then his explosion of energy as he bounds into the forest. This interaction pattern plays out in domestic forms, too. Imagine coming home from work with a friend, and as you enter your house your beloved dog runs across the room and greets you but not your friend with a boundless welcome. Your dog recognizes you, and you recognize your dog. In chapter 1, I wrote that if we go to a zoo we can sometimes see a child, or even adult, throwing food or a pebble at an animal, such as a lion, leopard, or great ape. The person is trying to get the animal's attention. I think throwing food or pebbles at a caged wild animal is a perversion of this deep need that people have to enact the wild form of this interaction pattern. It is such a deep need that even zoo signage, and zoo docents on the lookout, cannot fully stop zoo visitors from enacting this perversion.

So far, I have been speaking of interaction patterns as separate entities, but they usually coexist with many other interaction patterns, and dynamically bring forth many instantiations and experiences. As a rough analogy, think of an interaction pattern as a word in a language. Words have definitions. Words can be isolated. But words rarely exist by themselves. Usually words are combined to convey meaningful information and to bring forth ideas. Words can be combined in an endless number of ways. Similarly, although it is useful to generate individual interaction patterns to help gain clarity of their forms, we should not lose sight of how the interaction patterns mix and can topple on one another in our lived lives with nature.

Let us imagine that within five years we can establish 150 or 250 valid interaction patterns. They could then be used as psychological benchmarks to put technological nature "to the test." Using just the possible interaction patterns discussed so far, we could ask the following research questions: When interacting with a robot animal, can people enact the interaction pattern of recognition by a nonhuman other? Or does that recognition require that we attribute consciousness to the biological animal in a way that we do not to the robot animal? When taking a virtual nature walk through a DVD movie of a desert or mountain environment, can people enact the pattern of movement away from human settlement—and the return? Can people enact the pattern of finding respite from nature using nature in a virtual environment when the harshness of nature is not fully sensed or experienced? Can people enact the pattern of checks and balances by nature in a technological nature environment that, even if it checks us through its programming, still lies within the auspices of human control? The questions are many, and perhaps slightly "rigged" because the patterns deeply involve multisensorial interactions with a physical world that are difficult to even approximate technologically. But that is the point. If technological nature is a poor approximation of actual nature, and if that leads to our enacting only some or incomplete interaction patterns, we should know that limitation, and know it well.

Conclusion

We began with a seemingly simple question: How will we as a species adapt to technological nature? The question is of central importance because people often look at environmental destruction and technological developments and say something like "We'll adapt, don't worry, we'll be fine." I have argued in this chapter we will adapt, but that we should still worry because it is likely that we will not be fine.

My argument had the following structure. (i) Adaptation can be beneficial to a biological or social system, but it can also be harmful (or neutral). (ii) A system well adapted to its environment gains no moral claim. (iii) Adaptation refers broadly to a system's adjustment to its environment and occurs through a wide range of overlapping processes and mechanisms, which include homeostasis, acclimatization, addiction, habituation, equilibration, assimilation, and accommodation, along with the more standard account of biological evolution. (iv) In assessing whether an adaptation benefits or harms a system, the time frame must be specified, for what appears as a benefit (or harm) over a short time frame can become a harm (or benefit) over a longer time frame. (v) Five time frames are useful in assessing the benefits or harms to an adapting system: Level 1 harms last from milliseconds to one month; Level 2 harms last between 1 month and a lifetime; Level 3 harms last between 1 and 10 generations; Level 4 harms last between 100 and 500 generations; and Level 5 harms last over 500 generations. (vi) It is possible that given enough time (e.g., 500 or 5,000 generations) that some harms can disappear because a species changes into another species with different needs for its well-being, but it is virtually impossible to specify a priori for humans which harms can disappear, or how long it would take. (vii) To assess the benefits or harms of technological nature, and of human adaptation to increasingly sophisticated forms of technological nature in the years to come, we need to account for not only direct harms, but the less tractable harms of unfulfilled flourishing. (viii) A *nature language* has the potential to give voice to both types of harms, especially harms of unfulfilled flourishing. And (ix) the "vocabulary" of a nature language, *interaction patterns*, can be used as benchmarks for assessing the adequacy of current forms of technological nature, and for designing more successful forms in the years to come.

I do not accept the argument that we can simply adapt our way out of our problems. I granted that maybe in 10,000 years or more, it is unknown what our species will look like, and whether all the problems I speak of, in terms of the direct harms and the harms of unfulfilled flourishing, will exist in the way I have cast them here. I think many will. But even that does not matter, because we should not be making choices now based on harms that may or may not disappear 10,000 or 100,000 years into the future. There would be too much human suffering and harms of unfulfilled flourishing in the years between now and then to justify that choice.

Over a hundred years ago, in 1909, E. M. Forster wrote a short story, "The Machine Stops," where he described a future time when people live underground, detached from the natural world, and connected to one another only through the Machine. Forster's conception of the Machine was as an omnipotent, global technological entity that moderated and provided for all bodily and emotional needs of human beings. People became dependent on the Machine and learned to fear and often despise nature. In one early scene, Forster has a mother talking with her son by videoconference. The son says that he yearns for the actual experience of his mother and not the technologically mediated encounter. He tells his mother: "Men made it [the Machine], do not forget that. Great men, but men. The Machine is much, but it is not everything. I see something like you in this plate, but I do not see you. That is why I want you to come. Pay me a visit, so that we can meet face to face, and talk about the hopes that are in my mind" (Forster 1997, p. 140). Along similar lines, Forster writes: "He [the son] broke off, and she [the mother] fancied that he looked sad. She could not be sure, for the Machine did not transmit nuances of expression. It only gave a general idea of people—an idea that was good enough for all practical purposes" (p. 141). Through this story, Forster shows how the technological experience becomes "good" in the sense of "good enough" for basic functioning rather than good based on deeper capacities for humans to experience and to flourish.

We are a technological species. We will continue to design and build technological nature. If designed well, technological nature will offer substantive nature-like experiences. But I believe that technological nature, like Forster's Machine, will always result in a diminished experience compared to its natural counterpart. If that is true—and the results of my collaborative psychological research described in this book support that proposition—then we should employ technological nature as a bonus on actual nature, not as its substitute. Otherwise, we will come to believe, as we have to some degree already, that the "good enough" is the "good."

References

Adaptation. N.d. In *Merriam-Webster Online Dictionary*. Retrieved September 8, 2009, from <http://www.merriam-webster.com>.

Addiction. N.d. In *Merriam-Webster's Dictionary of Law*. Retrieved September 8, 2009, from <http://dictionary.reference.com>.

Alexander, C. 1979. *The Timeless Way of Building*. New York: Oxford University Press.

Alexander, C., S. Ishikawa, M. Silverstein, M. Jacobson, I. Fiksdahl-King, and S. Angel. 1977. *A Pattern Language*. New York: Oxford University Press.

American Museum of Natural History. 2008. Passenger pigeons. Retrieved October 1, 2008, from <http://www.amnh.org/exhibitions/expeditions/treasure_fossil/Treasures/Passenger_Pigeons/pigeons.html?dinos>.

Appadurai, A., ed. 1988. *The Social Life of Things: Commodities in Cultural Perspective*. New York: Cambridge University Press.

Balling, J. D., and J. H. Falk. 1982. Development of visual preference for natural environments. *Environment and Behavior* 14 (1):5–28.

Beck, A. M., and A. H. Katcher. 1996. *Between Pets and People: The Importance of Animal Companionship*. West Lafayette, IN: Purdue University Press.

Black death. N.d. In Wikipedia. Retrieved September 8, 2009, from <http://en.wikipedia.org/wiki/Black_Death>.

Borning, A., B. Friedman, and P. H. Kahn, Jr. 2004. Designing for human values in an urban simulation system: Value Sensitive Design and Participatory Design. Paper presented at the Participatory Design Conference, July, Toronto, Canada.

Boyle, M., C. Edwards, and S. Greenberg. 2000. The effects of filtered video on awareness and privacy. In *Proceedings of the Conference on Computer-Supported Collaborative Work*. New York: ACM Press, 1–10.

Bryant, B. 1985. The neighborhood walk: Sources of support in middle childhood. *Monographs of the Society for Research in Child Development* 50:1–114.

Buber, M. 1996. *I and Thou*. Trans. W. A. Kaufmann. New York: Touchstone.

Cole, M. 1991. Discussant to B. Friedman and P. H. Kahn, Jr. (Presenters), Who is responsible for what? And can what be responsible? The psychological boundaries of moral responsibility. Paper presented at the Society for Research in Child Development Biennial Meeting, April, Seattle, WA.

Damon, W. 1977. *The Social World of the Child.* San Francisco, CA: Jossey-Bass.

Dautenhahn, K., and I. Werry. 2004. Towards interactive robots in autism therapy. *Pragmatics & Cognition* 12:1–35.

Dautenhahn, K., I. Werry, T. Salter, and R. t. Boekhorst. 2003. Towards adaptive autonomous robots in autism therapy: Varieties of interactions. In *Proceedings of the IEEE International Symposium on Computational Intelligence in Robotics and Automation.* Piscataway, NJ: IEEE, 577–582.

Dawkins, R. 1976. *The Selfish Gene.* New York: Oxford University Press.

Dawson, G., and K. Toth. 2006. Autism spectrum disorders. In *Developmental Psychopathology: Risk, Disorder, and Adaptation,* vol. 3, ed. D. Cicchetti and D. J. Cohen. Hoboken, NJ: Wiley, 317–357.

Dean, B. 1997. A multitude of witnesses. *Northern Lights* 13:14–17.

DeVries, R., and B. Zan. 1994. *Moral Classrooms, Moral Children: Creating a Constructivist Atmosphere in Early Education.* New York: Teachers College Press.

Diamond, J. M. 1997. *Guns, Germs, and Steel: The Fates of Human Societies.* New York: Norton.

Diamond, J. M. 2005. *Collapse: How Societies Choose to Fail or Succeed.* New York: Penguin.

Dick, P. K. 1968. *Do Androids Dream of Electric Sheep?* London: Orion.

Dubos, R. [1965] 1980. *Man Adapting.* New Haven, CT: Yale University Press.

Ehrlich, P. R., and A. H. Ehrlich. 2008. *The Dominant Animal: Human Evolution and the Environment.* Washington, D.C.: Island Press.

Evans, G. W., S. V. Jacobs, and N. Frager. 1982. Adaptation to air pollution. *Journal of Environmental Psychology* 2:99–108.

Faber, M. 2005. The eyes of the soul. In *Vanilla Bright Like Eminem.* Orlando, FL: Harcourt, 37–48.

Farnham, S., L. Cheng, L. Stone, M. Zaner-Godsey, C. Hibbeln, K. Syrjala, et al. 2002. HutchWorld: Clinical study of computer-mediated social support for cancer patients and their caregivers. In *Proceedings of the Conference on Human Factors in Computer Systems.* New York: ACM Press, 375–382.

First Issue. 1993. *The Wilderness Society,* 56(200): 6.

Fischer, C. S. 1994. Widespread likings: Review of *The Biophilia Hypothesis. Science* 263:1161–1162.

Flenley, J., and P. Bahn. 2003. *The Enigmas of Easter Island.* New York: Oxford University Press.

Forster, E. M. 1997. The machine stops. In *The Science Fiction Century*, ed. D. G. Hartwell. New York: Tom Doherty, 139–160.

François, D., D. Polani, and K. Dautenhahn. 2007. On-line behaviour classification and adaptation to human–robot interaction styles. In *Proceedings of the ACM/IEEE International Conference on Human-Robot Interaction*. New York: ACM Press, 295–302.

Fredston, J. A. 2001. *Rowing to Latitude: Journeys along the Arctic's Edge*. New York: North Point Press.

Freud, S. 1930. *Civilization and Its Discontents*. Standard Edition 21, 59–145.

Friedman, B., A. Borning, J. L. Davis, B. T. Gill, P. H. Kahn, Jr., T. Kriplean, et al. 2008. Laying the foundations for public participation and value advocacy: Interaction design for a large scale urban simulation. In *Proceedings of the 9th Annual International Conference on Digital Government Research*. Stroudsburg, PA: Digital Government Society, 305–314.

Friedman, B., N. G. Freier, P. H. Kahn, Jr., P. Lin, and R. Sodeman. 2008. Office window of the future?—Field-based analyses of a new use of a large display. *International Journal of Human-Computer Studies* 66:452–465.

Friedman, B., D. C. Howe, and E. Felten. 2002. Informed consent in the Mozilla browser: Implementing value-sensitive design. In *Proceedings of the 35th Annual Hawaii International Conference on System Sciences*, 247.

Friedman, B., and P. H. Kahn, Jr. 2008. Human values, ethics, and design. In *The Human–Computer Interaction Handbook: Fundamentals, Evolving Technologies, and Emerging Applications*, rev. ed., ed. J. A. Jacko and A. Sears. Mahwah, NJ: Erlbaum, 1241–1266.

Friedman, B., P. H. Kahn, Jr., and A. Borning. 2006. Value Sensitive Design and information systems. In *Human–Computer Interaction and Management Information Systems: Foundations*, ed. P. Zhang and D. Galetta. Armonk, NY: M.E. Sharpe, 348–372.

Friedman, B., P. H. Kahn, Jr., and J. Hagman. 2003. Hardware companions?—What online AIBO discussion forums reveal about the human-robotic relationship. In *Proceedings of the Conference on Human Factors in Computer Systems*. New York: ACM Press, 273–280.

Friedman, B., P. H. Kahn, Jr., J. Hagman, R. L. Severson, and B. Gill. 2006. The watcher and the watched: Social judgments about privacy in a public place. *Human–Computer Interaction* 21:235–272.

Friedman, B., P. H. Kahn, Jr., and D. C. Howe. 2000. Trust online. *Communications of the ACM* 43 (12):34–40.

Friedman, B., L. Millett, and E. Felten. 2000. *Informed consent online: A conceptual model and design principles* (Tech. Rep. No. UW-CSE-00-12-02). Seattle: University of Washington, Computer Science & Engineering.

Friedman, B., I. E. Smith, P. H. Kahn, Jr., S. Consolvo, and J. Selawski. 2006. Development of a privacy addendum for open source licenses: Value sensitive design in industry. In *UbiComp 2006: Ubiquitous Computing*, ed. P. Dourish and A. Friday. Verlag, Germany: Springer, 194–211.

Fromm, E. 1964. *The Heart of Man*. New York: Harper & Row.

Furman, W. 1989. The development of children's social networks. In *Children's Social Networks and Social Supports*, ed. D. Belle. New York: Wiley, 151–172.

Gelman, R., E. Spelke, and E. Meck. 1983. What preschoolers know about animate and inanimate objects. In *The Acquisitions of Symbolic Skills*, ed. D. Rogers and J. A. Sloboda. New York: Plenum, 297–326.

Gelman, S. A., and E. M. Markman. 1986. Categories and induction in young children. *Cognition* 23:183–209.

Gilliam, J. E. 1995. *The Gilliam Autism Rating Scale*. Austin, TX: Pro-Ed.

Goldberg, K., ed. 2000. *The Robot in the Garden: Telerobotics and Telepistemology in the Age of the Internet*. Cambridge, MA: MIT Press.

Goldberg, K. 2005. Demonstrate: Tutorial. Retrieved September 7, 2009, from <http://demonstrate.berkeley.edu/tutorial.php>.

Gray wolf. N.d. In *Wikipedia*. Retrieved September 8, 2009, from <http://en.wikipedia.org/wiki/Gray_Wolf>.

Habituation. (n.d.). In *Merriam-Webster's Medical Dictionary*. Retrieved September 8, 2009, from <http://www.merriam-webster.com>.

Hand, G. 1997. The forest of forgetting. *Northern Lights* 13:11–13.

Helwig, C. C. 2002. Is it ever ok to exclude on the basis of race or gender? The role of context, stereotypes, and historical change. *Monographs of the Society for Research in Child Development* 67 (4):119–128.

Houston, J. A. 1995. *Confessions of an Igloo Dweller*. New York: Houghton Mifflin.

Hume, D. [1751] 1983. *An Enquiry Concerning the Principles of Morals*. Ed. J. B. Schneewind. Indianapolis, IN: Hackett.

Ingalls, R. P., and L. A. Perez, Jr. 2003. *Tampa Cigar Workers: A Pictorial History*. Gainesville: University Press of Florida.

Jancke, G., G. D. Venolia, J. Grudin, J. J. Cadiz, and A. Gupta. 2001. Linking public spaces: Technical and social issues. In *Proceedings of the Conference on Human Factors in Computer Systems*. New York: ACM Press, 530–537.

Joye, Y. 2007. Architectural lessons from environmental psychology: The case of biophilic architecture. *Review of General Psychology* 11 (4):305–328.

Kahn, P. H., Jr. 1999. *The Human Relationship with Nature: Development and Culture*. Cambridge, MA: MIT Press.

Kahn, P. H., Jr. 2002. Children's affiliations with nature: Structure, development, and the problem of environmental generational amnesia. In *Children and Nature: Psychological, Sociocultural, and Evolutionary Investigations*, ed. P. H. Kahn, Jr., and S. R. Kellert. Cambridge, MA: MIT Press, 93–116.

Kahn, P. H., Jr., and B. Friedman. 1995. Environmental views and values of children in an inner-city black community. *Child Development* 66: 1403–1417.

Kahn, P. H., Jr., B. Friedman, and I. S. Alexander. 2005. *Coding manual for "The distant gardener: What conversations in the Telegarden reveal about human–telerobotic interaction"* (Tech. Rep. No. IS-TR-2005-06-01). Seattle: University of Washington, Information School. Retrieved September 7, 2009, from <https://digital.lib.washington.edu/dspace/handle/1773/206>.

Kahn, P. H., Jr., B. Friedman, I. S. Alexander, N. G. Freier, and S. L. Collett. 2005. The distant gardener: What conversations in the Telegarden reveal about human–telerobotic interaction. In *Proceedings of the IEEE International Workshop on Robot and Human Interactive Communication*. Piscataway, NJ: IEEE, 13–18.

Kahn, P. H., Jr., B. Friedman, N. G. Freier, and R. Severson. 2003. *Coding manual for children's interactions with AIBO, the robotic dog—The preschool study* (Tech. Rep. No. UW-CSE-03-04-03). Seattle: University of Washington, Computer Science & Engineering.

Kahn, P. H., Jr., B. Friedman, B. Gill, J. Hagman, R. L. Severson, N. G. Freier, et al. 2008. A plasma display window?—The shifting baseline problem in a technologically-mediated natural world. *Journal of Environmental Psychology* 28:192–199.

Kahn, P. H., Jr., B. Friedman, and J. Hagman. 2002. "I care about him as a pal": Conceptions of robotic pets in online AIBO discussion forums. In *Proceedings of the Conference on Human Factors in Computer Systems* (extended abstracts). New York: ACM Press, 632–633.

Kahn, P. H., Jr., B. Friedman, D. R. Pérez-Granados, and N. G. Freier. 2006. Robotic pets in the lives of preschool children. *Interaction Studies: Social Behaviour and Communication in Biological and Artificial Systems* 7:405–436.

Kahn, P. H., Jr., B. Friedman, R. L. Severson, and E. N. Feldman. 2005. *Creativity tasks and coding system—Used in the plasma display window study* (Tech. Rep. No. IS-TR-2005-04-01). Seattle: University of Washington, Information School.

Kahn, P. H., Jr., H. Ishiguro, B. Friedman, T. Kanda, N. G. Freier, R. L. Severson, et al. 2007. What is a human? Toward psychological benchmarks in the field of human-robot interaction. *Interaction Studies: Social Behaviour and Communication in Biological and Artificial Systems* 8:363–390.

Kahn, P. H., Jr., and O. Lourenço. 2002. Water, air, fire, and earth: A developmental study in Portugal of environmental moral reasoning. *Environment and Behavior* 34:405–430.

Kahn, P. H., Jr., J. H. Ruckert, A. L. Reichert, and P. H. Hasbach. Forthcoming. A nature language. In *The Rediscovery of the Wild*, ed. P. H. Kahn, Jr., P. H. Hasbach, and J. H. Ruckert. Cambridge, MA: MIT Press.

Kahn, P. H., Jr., J. H. Ruckert, R. L. Severson, A. L. Reichert, and E. Fowler. 2010. A nature language: An agenda to catalog, save, and recover patterns of human–nature interaction. *Ecopsychology* 2:59–66.

Kahn, P. H., Jr., R. L. Severson, and J. H. Ruckert. 2009. The human relation with nature and technological nature. *Current Directions in Psychological Science* 18:37–42.

Kalai, A., and M. Siegel. 1998. Improved rendering of parallax panoramagrams for a time-multiplexed autostereoscopic display. In *Proceedings of SPIE, 3295,* ed. M. T. Bolas, S. S. Fisher, and J. O. Merritt. Bellingham, WA: SPIE, 211–217.

Kaplan, R., and S. Kaplan. 1989. *The Experience of Nature: A Psychological Perspective.* New York: Cambridge University Press.

Kaplan, R., S. Kaplan, and R. L. Ryan. 1998. *With People in Mind: Design and Management of Everyday Nature.* Washington, D.C.: Island Press.

Kaplan, F., P.-Y. Oudeyer, E. Kubinyi, and A. Miklosi. 2002. Robotic clicker training. *Robotics and Autonomous Systems* 38:197–206.

Katcher, A., and G. Wilkins. 1993. Dialogue with animals: Its nature and culture. In *The Biophilia Hypothesis*, ed. S. R. Kellert and E. O. Wilson. Washington, D.C.: Island Press, 173–197.

Kellert, S. R. 2005. *Building for Life: Designing and Understanding the Human– Nature Connection.* Washington, D.C.: Island Press.

Kellert, S. R., J. H. Heerwagen, and M. L. Mador, eds. 2008. *Biophilic Design.* Hoboken, NJ: Wiley.

Kellert, S. R., and E. O. Wilson, eds. 1993. *The Biophilia Hypothesis.* Washington, D.C.: Island Press.

Kiesler, S., ed. 1997. *Culture of the Internet.* Mahwah, NJ: Erlbaum.

Killen, M. 1989. Context, conflict, and coordination in early social development. In *Social Interaction and the Development of Children's Understanding*, ed. L. T. Winegar. Norwood, NJ: Ablex, 119–146.

Killen, M. 2007. Children's social and moral reasoning about exclusion. *Current Directions in Psychological Science* 16:32–36.

Koegel, R. L., L. K. Koegel, and E. K. McNerney. 2001. Pivotal areas in intervention for autism. *Journal of Clinical Child Psychology* 30:19–32.

Kohak, E. 1984. *The Embers and the Stars: A Philosophical Inquiry into the Moral Sense of Nature.* Chicago: University of Chicago Press.

Kohlberg, L., and R. Mayer. 1972. Development as the aim of education. *Harvard Educational Review* 42:449–496.

Kozima, H., C. Nakagawa, N. Kawai, D. Kosugi, and Y. Yano. 2004. A humanoid in company with children. In *Proceedings of the IEEE/RAS International Conference on Humanoid Robots.* Piscataway, NJ: IEEE, 470–477.

Kozima, H., C. Nakagawa, and Y. Yasuda. 2005. Designing and observing human–robot interactions for the study of social development and its disorders.

In *Proceedings of the IEEE International Symposium on Computational Intelligence in Robotics and Automation.* Piscataway, NJ: IEEE, 41–46.

Kozima, H., and H. Yano. 2001. Designing a robot for contingency-detection game. Paper presented at the Workshop on Robotic and Virtual Interactive Systems in Autism Therapy, September, Hatfield, Hertfordshire, UK.

Krathwohl, D. R. 1998. *Methods of Educational and Social Science Research: An Integrated Approach*, 2nd ed. Reading, MA: Addison-Wesley.

Küller, R., and C. Lindsten. 1992. Health and behavior of children in classrooms with and without windows. *Journal of Environmental Psychology* 12:305–317.

Küller, R., and L. Wetterberg. 1993. Melatonin, cortisol, EEG, ECG, and subjective comfort in healthy humans: Impact of two fluorescent lamp types at two light intensities. *Lighting Research & Technology* 25 (2):71–81.

Lakoff, G. 1987. *Women, Fire, and Dangerous Things: What Categories Reveal about the Mind.* Chicago: University of Chicago Press.

Latour, B. 1992. Where are the missing masses? The sociology of a few mundane artifacts. In *Shaping Technology/Building Society: Studies in Sociotechnical Change*, ed. W. E. Bijker and J. Law. Cambridge, MA: MIT Press, 225–258.

Lawrence, E. A. 1993. The sacred bee, the filthy pig, and the bat out of hell: Animal symbolism as cognitive biophilia. In *The Biophilia Hypothesis*, ed. S. R. Kellert and E. O. Wilson. Washington, D.C.: Island Press, 301–341.

Leather, P., M. Pyrgas, D. Beale, and C. Lawrence. 1998. Windows in the workplace: Sunlight, view, and occupational stress. *Environment and Behavior* 30:739–762.

Liu, C., K. Conn, N. Sarkar, and W. Stone. 2007. Affect recognition in robot assisted rehabilitation of children with autism spectrum disorder. In *Proceedings of the IEEE International Conference on Robotics and Automation.* Piscataway, NJ: IEEE, 1755–1760.

Louv, R. 2005. *Last Child in the Woods: Saving Our Children from Nature-Deficit Disorder.* Chapel Hill, NC: Algonquin.

MacDorman, K. F. 2005. Androids as an experimental apparatus: Why is there an uncanny valley and can we exploit it? In *Proceedings of the CogSci 2005 Workshop: Toward Social Mechanisms of Android Science.* Osaka, Japan: Osaka University, 108–118.

MacDorman, K. F., and H. Ishiguro. 2006. The uncanny advantage of using androids in cognitive and social science research. *Interaction Studies: Social Behaviour and Communication in Biological and Artificial Systems* 7:297–337.

Mammoth Cave National Park. N.d. Retrieved July 13, 2010, from <http://www.nps.gov/maca/forteachers/planafieldtrip2.htm>.

Marshall, L. 1976. *The !Kung of Nyae Nyae.* Cambridge, MA: Harvard University Press.

Martin, F., and J. Farnum. 2002. Animal-assisted therapy for children with pervasive developmental disorders. *Western Journal of Nursing Research* 24:657–670.

McGrath, J. E. 1995. Methodology matters: Doing research in the behavioral and social sciences. In *Readings in Human–Computer Interaction: Toward the Year 2000*, 3rd ed., ed. R. M. Baecker, J. Grudin, W. A. S. Buxton, and S. Greenberg. San Francisco, CA: Morgan Kaufmann, 152–169.

McKenna, K. Y. A., and J. A. Bargh. 2000. Plan 9 from cyberspace: The implications of the Internet for personality and social psychology. *Personality and Social Psychology Review* 4:57–75.

Meloy, E. 1997. Waiting its occasions. *Northern Lights* 13:4–6.

Melson, G. F. 2001. *Why the Wild Things Are: Animals in the Lives of Children*. Cambridge, MA: Harvard University Press.

Melson, G. F., P. H. Kahn, Jr., A. Beck, and B. Friedman. 2009. Robotic pets in human lives: Implications for the human-animal bond and for human relationships with personified technologies. *Journal of Social Issues* 65: 545–567.

Melson, G. F., P. H. Kahn, Jr., A. Beck, B. Friedman, T. Roberts, E. Garrett, et al. 2009. Children's behavior toward and understanding of robotic and living dogs. *Journal of Applied Developmental Psychology* 30:92–102.

Millett, L., B. Friedman, and E. Felten. 2001. Cookies and Web browser design: Toward realizing informed consent online. In *Proceedings of the Conference on Human Factors in Computer Systems*. New York: ACM Press, 46–52.

Milton, J. [1674] 1978. Paradise lost. In *John Milton: Complete Poems and Major Prose*, ed. M. Y. Hughes. Indianapolis, IN: Odyssey Press, 211–469.

Montagner, H., A. Restoin, D. Rodriguez, V. Ullmann, M. Viala, D. Laurent, et al. 1988. Social interactions of young children with peers and their modifications in relation to environmental factors. In *Social Fabrics of the Mind*, ed. M. R. A. Chance and D. R. Omark. Hillsdale, NJ: Erlbaum, 237–259.

Moore, G. E. [1903] 1978. *Principia Ethica*. Cambridge: Cambridge University Press.

Moulton, G. E., ed. 1993. *The Definitive Journals of Lewis & Clark: Over the Rockies to St. Louis*, vol. 8. Lincoln: University of Nebraska Press.

Mumford, L. 1961. *The City in History*. New York: Harcourt.

Myers, G. 1998. *Children and Animals: Social Development and Our Connections to Other Species*. Boulder, CO: Westview Press.

Myers, O. E., Jr. 2007. *The Significance of Children and Animals: Social Development and Our Connection to Other Species*, rev. ed. West Lafayette, IN: Purdue University Press.

Mynatt, E. D., A. Adler, M. Ito, C. Linde, and V. L. O'Day. 1999. The network communities of SeniorNet. In *Proceedings of the Sixth European Conference on Computer Supported Cooperative Work*, ed. S. Bodker, M. Kyng, and K. Schmidt. Dordrecht, Netherlands: Kluwer, 219–238.

Nass, C., Y. Moon, J. Morkes, E. Kim, and B. J. Fogg. 1997. Computers are social actors: A review of current research. In *Human Values and the Design of Computer Technology*, ed. B. Friedman. New York: Cambridge University Press, 137–162.

Nucci, L. P. 1981. The development of personal concepts: A domain distinct from moral and societal concepts. *Child Development* 52:114–121.

Nucci, L. P. 1996. Morality and the personal sphere of actions. In *Values and Knowledge*, ed. E. S. Reed, E. Turiel, and T. Brown. Mahwah, NJ: Erlbaum, 41–60.

Orians, G. H., and J. H. Heerwagen. 1992. Evolved responses to landscapes. In *The Adapted Mind: Evolutionary Psychology and the Generation of Culture*, ed. J. H. Barkow, L. Cosmides, and J. Tooby. New York: Oxford University Press, 555–579.

Orlikowski, W. J. 2000. Using technology and constituting structures: A practice lens for studying technology in organizations. *Organization Science* 11: 404–428.

Orr, D. W. 1993. Love it or lose it: The coming biophilia revolution. In *The Biophilia Hypothesis*, ed. S. R. Kellert and E. O. Wilson. Washington, D.C.: Island Press, 415–440.

Owens, M., and D. Owens. 2006. *Secrets of the Savanna: Twenty-three Years in the African Wilderness Unraveling the Mysteries of Elephants and People*. New York: Houghton Mifflin.

Papworth, S. K., J. Rist, L. Coad, and E. J. Milner-Gulland. 2009. Evidence for shifting baseline syndrome in conservation. *Conversation Letters* 2:93–100.

Pauly, D. 1995. Anecdotes and the shifting baseline syndrome of fisheries. *Trends in Ecology & Evolution* 10:430.

Pelto, P. J., and G. H. Pelto. 1978. *Anthropological Research: The Structure of Inquiry*. New York: Cambridge University Press.

Piaget, J. 1983. Piaget's theory. In *Handbook of Child Psychology*, vol. 1: *History, Theory, and Methods*, 4th ed., series ed. P. H. Mussen, vol. ed. W. Kessen. New York: Wiley, 103–128.

Pommer, E. (producer), and F. Lang (director). 1927. *Metropolis*. [Motion picture.] Potsdam, Germany: Babelsberg.

Preece, J. 1998. Empathic communities: Reaching out across the Web. *Interactions* 5(2):32–43.

Preece, J. 1999. Empathic communities: Balancing emotional and factual communication. *Interacting with Computers* 12:63–77.

Price, R., R. A. Lovka, and B. Lovka. 2000. *Classic Droodles*. Beverly Hills, CA: Tallfellow Press.

Pyle, R. M. 2002. Eden in a vacant lot: Special places, species, and kids in the neighborhood of life. In *Children and Nature: Psychological, Sociocultural, and Evolutionary Investigations*, ed. P. H. Kahn, Jr., and S. R. Kellert. Cambridge, MA: MIT Press, 305–327.

Rabb, G. B. 2004. The evolution of zoos from menageries to centers of conservation and caring. *Curator* 47:237–246.

Radikovic, A. S., J. J. Leggett, R. S. Ulrich, and J. Keyser. 2005. Artificial window view of nature. In *Proceedings of the Conference on Human Factors in Computer Systems*. New York: ACM Press, 1993–1996.

Reeves, B., and C. Nass. 1996. *The Media Equation: How People Treat Computers, Television, and New Media Like Real People and Places*. New York: Cambridge University Press.

Rheingold, H. 1993. *The Virtual Community*. Reading, MA: Addison-Wesley.

Robins, B., and K. Dautenhahn. 2006. The role of the experimenter in HRI research—A case study evaluation of children with autism interacting with a robotic toy. In *Proceedings of the IEEE International Symposium on Robot and Human Interactive Communication*. Piscataway, NJ: IEEE, 646–651.

Robins, B., K. Dautenhahn, R. t. Boekhorst, and A. Billard. 2004a. Effects of repeated exposure to a humanoid robot on children with autism. In *Designing a More Inclusive World*, ed. S. Keates, J. Clarkson, P. Langdon, and P. Robinson. London: Springer, 225–236.

Robins, B., K. Dautenhahn, R. t. Boekhorst, and A. Billard. 2004b. Robots as assistive technology—Does appearance matter? In *Proceedings of the IEEE International Workshop on Robot and Human Interactive Communication*. Piscataway, NJ: IEEE, 277–282.

Robins, B., K. Dautenhahn, R. t. Boekhorst, and A. Billard. 2005. Robotic assistants in therapy and education of children with autism: Can a small humanoid robot help encourage social interaction skills? *Universal Access in the Information Society* 4:105–120.

Robins, B., K. Dautenhahn, and J. Dubowski. 2005. Robots as isolators or mediators for children with autism? A cautionary tale. In *Proceedings of the Symposium on Robot Companions: Hard Problems and Open Challenges in Robot-Human Interaction, AISB'05 Convention*. Brighton, East Sussex, UK: SSAISB, 82–88.

Robins, B., P. Dickerson, P. Stribling, and K. Dautenhahn. 2004. Robot-mediated joint attention in children with autism. *Interaction Studies: Social Behaviour and Communication in Biological and Artificial Systems* 5:161–198.

Rogers, S. J. 2006. Evidence-based interventions for language development in young children with autism. In *Social and Communication Development in Autism Spectrum Disorders: Early Identification, Diagnosis, and Intervention*, ed. T. Charman and W. Stone. New York: Guilford Press, 143–179.

Rolston, H., III. 1989. *Philosophy Gone Wild*. Buffalo, NY: Prometheus Books.

Ruckert, J. H., and P. H. Kahn, Jr. 2007. Biophilia—Past conundrums and new directions. Paper presented at the Psychology-Ecology-Sustainability Conference, June, Portland, OR.

Ruse, M., and E. O. Wilson. 1985. The evolution of ethics. *New Scientist* 108 (17):50–52.

Sams, M. J., E. V. Fortney, and S. Willenbring. 2006. Occupational therapy incorporating animals for children with autism: A pilot investigation. *American Journal of Occupational Therapy* 60:268–274.

Scassellati, B. 2005. Quantitative metrics of social response for autism diagnosis. In *Proceedings of the IEEE International Workshop on Robot and Human Interactive Communication*. Piscataway, NJ: IEEE, 585–590.

Scassellati, B. 2007. How social robots will help us to diagnose, treat, and understand autism. In *Robotics Research: Results of the 12th International Symposium ISRR*, ed. S. Thrun, R. Brooks, and H. Durrant-Whyte. Heidelberg, Germany: Springer, 552–563.

Searle, J. R. 1990. Is the brain's mind a computer program? *Scientific American* 262 (1):26–31.

Shepard, P. 1998. *Coming Home to the Pleistocene*. Washington, D.C.: Island Press.

Shibata, T., K. Wada, T. Saito, and K. Tanie. 2005. Human interactive robot for psychological enrichment and therapy. In *Proceedings of the Symposium on Robot Companions: Hard Problems and Open Challenges in Robot-Human Interaction, AISB'05 Convention*. Brighton, East Sussex, UK: SSAISB, 98–107.

Siegler, R. S. 1996. *Emerging Minds: The Process of Change in Children's Thinking*. Oxford: Oxford University Press.

SkyV—From the Sky Factory. 2008. Retrieved November 15, 2008, from <http://www.theskyfactory.com/products/ceilings/SkyV>.

Smetana, J. G. 1995. Morality in context: Abstractions, ambiguities and applications. In *Annals of Child Development*, vol. 10, ed. R. Vasta. London: Kingsley, 83–130.

Smetana, J. G. 2006. Social domain theory: Consistencies and variations in children's moral and social judgments. In *Handbook of Moral Development*, ed. M. Killen and J. G. Smetana. Mahwah, NJ: Erlbaum, 119–154.

Smith, M. R., and L. Marx. 1994. *Does Technology Drive History? The Dilemma of Technological Determinism*. Cambridge, MA: MIT Press.

Somé, M. P. 1995. *Of Water and the Spirit: Ritual, Magic, and Initiation in the Life of an African Shaman*. New York: Penguin.

Stanton, C. M., P. H. Kahn, Jr., R. L. Severson, J. H. Ruckert, and B. T. Gill. 2008. Robotic animals might aid in the social development of children with autism. In *Proceedings of the ACM/IEEE International Conference on Human–Robot Interaction*. New York: ACM Press, 97–104.

Svensson, M., K. Hook, J. Laaksolahti, and A. Waern. 2001. Social navigation of food recipes. In *Proceedings of the Conference on Human Factors in Computer Systems*. New York: ACM Press, 341–348.

Swanson, J., D. Johnson, and M. V. Kamp. 2006. A critical review of the concepts of "Environmental Generational Amnesia" and "Nature Deficit Disorder." Unpublished manuscript.

Teale, E. W., ed. 1976. *The Wilderness World of John Muir*. Boston: Houghton Mifflin.

The Telegarden Web site. N.d. Retrieved September 4, 2009, from <http://goldberg.berkeley.edu/garden/Ars>.

Thomas, E. M. 1959. *The Harmless People*. New York: Knopf.

Thomas, E. M. 2006. *The Old Way: A Story of the First People*. New York: Farrar, Straus & Giroux.

Tilburg, J. A. V. 1994. *Easter Island: Archaeology, Ecology, and Culture*. Washington, D.C.: Smithsonian Institution Press.

Torrance, E. P. 1962. *Guiding Creative Talent*. Englewood Cliffs, NJ: Prentice Hall.

The tradition of the lector. 2004. *Encore Arts Programs* 24(2): 6. [Brochure.] Seattle, WA: Encore Media.

Turiel, E. 1983. *The Development of Social Knowledge*. New York: Cambridge University Press.

Turiel, E. 1998. The development of morality. In *Handbook of Child Psychology: Social, Emotional, and Personality Development*, 5th ed., vol. 3, ed. W. Damon and N. Eisenberg. New York: Wiley, 863–932.

Turiel, E. 2002. *The Culture of Morality: Social Development and Social Opposition*. Cambridge: Cambridge University Press.

Turiel, E., C. Hildebrandt, and C. Wainryb. 1991. Judging social issues: Difficulties, inconsistencies, and consistencies. *Monographs of the Society for Research in Child Development* 56 (2):1–103.

Turiel, E., M. Killen, and C. C. Helwig. 1987. Morality: Its structure, functions, and vagaries. In *The Emergence of Morality in Young Children*, ed. J. Kagan and S. Lamb. Chicago: University of Chicago Press, 155–244.

Turkle, S. 1997. *Life on the Screen: Identity in the Age of the Internet*. New York: Simon & Schuster.

Ulrich, R. S. 1984. View through a window may influence recovery from surgery. *Science* 224:420–421.

Ulrich, R. S. 1993. Biophilia, biophobia, and natural landscapes. In *The Biophilia Hypothesis*, ed. S. R. Kellert and E. O. Wilson. Washington, D.C.: Island Press, 73–137.

Ulrich, R. S., R. F. Simons, B. D. Losito, E. Fiorito, M. A. Miles, and M. Zelson. 1991. Stress recovery during exposure to natural and urban environments. *Journal of Environmental Psychology* 11:201–230.

U.S. Department of Health, Education, and Welfare. 1978. *The Belmont report: Ethical principles and guidelines for the protection of human subjects of research* (DHEW Publication No. [OS] 78-0012). Washington, D.C.: U.S. Government Printing Office.

Waddell, P. 2002. UrbanSim: Modeling urban development for land use, transportation and environmental planning. *Journal of the American Planning Association* 68:297–314.

Wellman, H., A. K. Hickling, and C. A. Schult. 2000. Young children's psychological, physical, and biological explanations. In *Childhood Cognitive Development: The Essential Readings*, ed. K. Lee. Malden, MA: Blackwell, 267–288.

Witherspoon, B., and M. Petrick. 2008. Scientific research and sky image ceilings. Unpublished manuscript.

Wilson, E. O. 1975. *Sociobiology: The New Synthesis*. Cambridge, MA: Harvard University Press.

Wilson, E. O. 1984. *Biophilia*. Cambridge, MA: Harvard University Press.

Wilson, E. O. 1992. *The Diversity of Life*. Cambridge, MA: Harvard University Press.

Wilson, E. O. 1993. Biophilia and the conservation ethic. In *The Biophilia Hypothesis*, ed. S. R. Kellert and E. O. Wilson. Washington, D.C.: Island Press, 31–41.

Wilson, E. O. 1994. *Naturalist*. Washington, D.C.: Island Press.

Wilson, E. O. 2002. *The Future of Life*. New York: Knopf.

Index